高等学校摄影测量与遥感系列教材

遥感技术基础双语讲义

Bilingual Textbook for Introduction to Remote Sensing

蔡国印　杜明义　编译

武汉大学出版社

图书在版编目(CIP)数据

遥感技术基础双语讲义.汉、英/蔡国印,杜明义编译.—武汉:武汉大学出版社,2016.5(2024.8重印)
高等学校摄影测量与遥感系列教材
ISBN 978-7-307-17649-2

Ⅰ.遥… Ⅱ.①蔡… ②杜… Ⅲ.遥感技术—双语教学—高等学校—教材—汉、英 Ⅳ.TP7

中国版本图书馆 CIP 数据核字(2016)第 040171 号

责任编辑:王金龙　　责任校对:李孟潇　　版式设计:马　佳

出版发行:武汉大学出版社　　(430072　武昌　珞珈山)
(电子邮箱:cbs22@whu.edu.cn 网址:www.wdp.com.cn)
印刷:湖北云景数字印刷有限公司
开本:787×1092　1/16　印张:14.25　字数:333 千字
版次:2016 年 5 月第 1 版　　2024 年 8 月第 2 次印刷
ISBN 978-7-307-17649-2　　定价:29.00 元

版权所有,不得翻印;凡购买我社的图书,如有质量问题,请与当地图书销售部门联系调换。

前　言

目前，许多领域都将环境遥感作为一个重要工具来进行测图、监测以及评估自然和人造景观。遥感数据可以从全球、国家、区域和地方等不同层次提供多尺度的信息。同时，遥感数据也包含了重要的时相和土地利用变化信息。经过处理后的遥感数据可以提供独特的信息，并能够与其他形式的数据(比如地理信息系统(GIS)所支持的格式)进行整合。因此，从事测绘、地理信息系统等专业的学生应该理解遥感数据的来源、分析工具、潜在的应用以及影像解译和影像产品等方面的知识。

遥感技术基础是遥感科学入门水平的课程，主要包括遥感的基本原理及其物理基础。本课程对遥感的基本概念进行了详细的介绍以便于学生能够将其应用于其他学科。

本课程以基于卫星的被动遥感系统为例对遥感技术进行了详细的讲解，同时基于现有的遥感软件，对遥感影像在地球资源应用方面的处理和分析进行了阐述。

本课程的主要目标为介绍遥感的理论、概念，使得学生能够掌握遥感的基本内容，能够处理一些常用的遥感数据并能够对处理结果进行理解和分析。本课程的学习可以培养学生将环境遥感技术应用于相关领域的意识。

本书的编辑出版得到了北京建筑大学测绘学院在经费上的大力支持，得到了讲授本课程同仁的中肯意见，武汉大学出版社王金龙分社长、王宵君编辑为本书的出版付出了辛勤的劳动，在此一并表示衷心的感谢。

Preface

Remote sensing of the environment is a major tool used by a range of disciplines to map, monitor and evaluate natural and built landscapes. Remotely sensed data can provide large and small scale information at the global, national, regional and local levels. It can also provide important temporal and land change information. Processed remotely sensed data can provide stand alone information and can be integrated with other forms of data, for example, within geographical information systems. Therefore, an understanding of the genesis of the information, analysis techniques, potential applications and interpretation of images and image-based products will be required in many situations for a range of disciplines.

This unit is an introductory level remote sensing unit; it covers the basic principles and includes discussions on the physical basis of remote sensing. This unit also covers the fundamental concepts underpinning remote sensing applications to a range of disciplines.

A typical satellite-based passive remote sensing system is explored in detail. Processing and analysis of remotely sensed satellite images is carried out from an earth resource perspective using geospatial image processing software.

Theoretical concepts that provide a basis for remote sensing, provide practical experience and understanding of a range of computer-based digital image processing techniques and assist students to develop an appreciation of the range of applications of remote sensing of environment.

This publication was funded by School of Geomatics and Urban Spatial Information in Beijing University of Civil Engineering and Architecture. The authors gratefully acknowledge on colleagues for their critical suggestions. Mostly, my thanks are also due to editors Wang Jinlong and Wang Xiaojun in Wuhan University Press for their efforts in revising and editing this textbook.

目　录

第1章　遥感概述 1
  1.1　定义 1
  1.2　遥感发展的里程碑 2
  1.3　遥感过程 3
  1.4　遥感的关键概念 5
  1.5　遥感的应用 10

第2章　电磁波谱 12
  2.1　电磁波 12
  2.2　适合遥感的波谱范围 15
  2.3　典型地物的光谱曲线 17
  2.4　大气对传感信号的影响 22
  2.5　大气辐射与地表的相互作用 27

第3章　遥感平台和传感器 30
  3.1　简介 30
  3.2　遥感平台 30
  3.3　常用遥感数据源 32
  3.4　航天器简介 38

第4章　影像数据获取 41
  4.1　光机扫描仪 42
  4.2　电荷耦合设备（CCD） 43
  4.3　扫描系统 43
  4.4　遥感数据 46

第5章　影像校正和预处理 49
  5.1　预处理 49
  5.2　几何校正 49
  5.3　辐射校正 54
  5.4　特征提取 57

5.5 图像裁剪·· 60

第6章 影像解译··· 62
6.1 简介·· 62
6.2 影像解译元素··· 63
6.3 影像解译策略··· 67

第7章 影像分类··· 69
7.1 引言·· 69
7.2 信息类和光谱类·· 71
7.3 非监督分类·· 72
7.4 监督分类··· 77
7.5 纹理分类··· 86
7.6 模糊分类··· 87
7.7 神经网络分类··· 89
7.8 数字影像分类的后处理··· 90

第8章 精度评定··· 93
8.1 参考数据··· 93
8.2 精度评定方法··· 94

Contents

Chapter 1 Overview of Remote Sensing 99
  1.1 Definitions 99
  1.2 Milestones in the History of Remote Sensing 100
  1.3 Remote Sensing Process 102
  1.4 Key Concepts of Remote Sensing 104
  1.5 Applications of RS 110

Chapter 2 Electromagnetic Spectrum 113
  2.1 Electromagnetic Waves 113
  2.2 The Spectral Region Used in RS 117
  2.3 Typical EMR Spectral 120
  2.4 Interactions with the Atmosphere 125
  2.5 Interactions with Surfaces 131

Chapter 3 Platforms and Sensors 134
  3.1 Introduction 134
  3.2 Platforms 134
  3.3 Remote Sensing Data Sources 136
  3.4 Mission types 143

Chapter 4 Acquiring Remote Sensing Data 147
  4.1 Optical-mechanical scanners 148
  4.2 Charge-coupled Device (CCD) 148
  4.3 Scanner Systems 150
  4.4 Remotely Sensed Data 153

Chapter 5 Image corrections and preprocessing 156
  5.1 Preprocessing 156
  5.2 Geometric Corrections 156
  5.3 Radiometric Correction 162
  5.4 Feature Extraction 165

    5.5   Subsets ·· 169

Chapter 6   Image Interpretation ··· 171
    6.1   Introduction ··· 171
    6.2   Elements of Image Interpretation ·· 173
    6.3   Image Interpretation Strategies ·· 176

Chapter 7   Classification of Remotely Sensed Data ·· 178
    7.1   Introduction ··· 178
    7.2   Information Classes and Spectral Classes ·· 181
    7.3   Unsupervised Classification ··· 182
    7.4   Supervised Classification ·· 188
    7.5   Textural Classifiers ··· 200
    7.6   Fuzzy Clustering ·· 202
    7.7   Artificial Neural Networks ·· 204
    7.8   Post Processing of Digital Classified Imagery ·· 205

Chapter 8   Accuracy Assessment ··· 208
    8.1   Reference Data ·· 208
    8.2   Accuracy Assessment Method ··· 210

Reference ·· 215

第1章 遥感概述

1.1 定义

前人从不同的角度对遥感的定义进行了阐述：

"遥感的定义很多，但是最基本的一点是通过非接触式探测地物的艺术或科学。"(Fischer et al., 1976, p. 34)

"遥感是通过非接触或联系地物的方式来实现地物物理数据的获取。"(Lintz & Simonett, 1976, p. 1)

"影像的获取是通过传感器而非传统的相机拍摄，比如，光扫描即利用相机或胶卷之外的辐射，如微波、雷达、热红外、紫外，并利用多光谱，特殊技术等来处理和解译遥感影像以用于生产传统的地图、专题图、资源调查等，在农业、考古、森林、地理、地质等方面广泛应用。"(美国摄影测量学会)

"遥感是利用设备在一定距离之外探测目标。"(Barrett & Curtis, 1976, p. 3)

"遥感一词广义上来说即为一定距离之外的侦查。"(Colwell, 1966, p. 71)

"遥感，虽然不能精确定义，但是其包括以一定距离获取地表图片或地磁记录的各种方法，以及图片数据的处理。因此，从广义上来说，遥感即为利用传感设备来记录和探测目标区域电磁辐射。这种辐射可能由目标物直接发出，也可能是反射的太阳辐射，或者是接收由传感器发出经目标物反射回的能量。"(White, 1977, p. 1-2)

"遥感目前被很多科学家用以研究遥远的目标，比如地球、月球以及其他星球表面和大气以及冰川现象等。广义来说，遥感将现代传感器、数据处理、设备、信息理论和处理方法、通信理论和设备、航空和航天飞机以及巨系统理论和时间等充分利用以实现地表空间的探测目标"。(国家科学院, 1970, p. 1)

"遥感是在一定距离之外非接触式获取地物信息的科学。目前的遥感系统探测地物用的最多的就是地物辐射的电磁能量。尽管有其他形式如地震波、声呐波和重力强度等，但是我们所关注的是电磁波。"(D. A. Landgrebe, In Swain and Davis, 1978)

对这些定义中共性问题的挖掘可以帮助我们发现遥感中最重要的特征。粗略地看一下这些定义，很容易得出：遥感即为一定距离之外的信息获取。这是一个关于遥感的广义定义，如果要针对某一个课程的学习，那么需要对其进行进一步的精练，以便于更好地掌握该课程的知识。

我们这里主要讨论由反射或发射的电磁波来探测陆地和水体的表面。因此，其他形式的遥感，比如对地磁、大气或者人体温度等的感知，在此不做讨论。此外，我们主要讨论成像的传感器，那些不成像的传感器，比如某类激光，在此不做讨论。他们虽然从属于遥感领域，但是简便起见，本课程不做讨论。

针对于我们要讨论的内容，遥感可以定义为：遥感即通过非接触式方式获取地表信息的科学(在一定程度上，也是艺术)。遥感是通过感应和记录反射或发射的能量来获取信息，同时对所获取的信息进行处理、分析和应用。

1.2 遥感发展的里程碑

遥感领域所涉及的范围可以通过追踪其核心概念的发展历史来进行阐述。下面几项关键事件的发展可以用来追述该领域的演变(表1.1)。

表1.1　　　　　　　　　　遥感发展历史上的里程碑事件

时间	事件
1800	William Herschel 发现了红外射线
1839	尝试进行照片的拍摄
1847	A. H. L. Fizeau & J. B. L. Foucault 展示了红外波谱，其与可见光有共同特性
1850—1860	利用气球进行拍摄
1973	James Clerk Maxwell 提出了电磁波理论
1909	尝试航拍
1914—1918	第一次世界大战中使用航空侦察
1920—1930	航拍和摄影测量的发展及其初步应用
1929—1939	经济衰退引起环境危机促使政府开始使用航空摄影测量
1930—1940	雷达开始在德国、美国和英国发展起来
1939—1945	第二次世界大战：电磁波谱不可见部分可以投入使用，同时开始进行航片数据获取和解译的培训
1950—1960	军事方面的研究和开发
1956	Colwell 开始研究利用红外照片进行植被病害的探测
1960—1970	首次使用"遥感"这一术语；TIROS 气象卫星发射；Skylab 从太空进行遥感对地观测
1972	Landsat 1 发射
1970—1980	数字影像处理技术得到迅速发展
1980—1990	Landsat 4：新一代陆地卫星探测传感器
1986	SPOT 法国地球观测卫星
1980s	高光谱传感器的发展
1990s	全球遥感系统

Evelyn Pruit，一位在海军海事研究办公室的科学家，当她发现利用"航空摄影"一词

已经不能用来精确描述利用可见光之外的波谱进行拍摄得到的许多形式的影像时，提出了"遥感"一词。早在20世纪60年代，美国国家宇航局(NASA)设立了一项遥感方面的研究项目，主要用于在下一个十年里，支持美国的研究机构在美国国家范围内从事遥感的研究。在此期间，美国国家科学院(NAS)的一个委员会研究了将遥感用于农业和森林领域的可能性。在1970年，NAS提交了其研究成果报告，指出遥感这一新出现的探测领域可以用于很多方面。

1972年，Landsat 1的升空是遥感发展的又一次里程碑事件。Landsat 1是众多探测地表的地球轨道卫星之一，它首次实现了对地系统地重复观测。每景Landat 1影像在几个电磁波谱区间对地成像，为其在诸多领域的应用提供中分辨率的数据。Landsat对遥感的重要意义至少可以概括为如下三点：首先，能够持续、稳定的提供区域地表的多光谱数据，这促进了大量的科研人员从事多光谱数据的分析工作。其次，Landsat促进了人们对遥感数据进行数字分析的迅速和全面扩展。再次，Landsat项目为世界上其他地球观测卫星的设计和运营树立了典范。

1.3 遥感过程

1.3.1 遥感过程概述

因为遥感是由许多个相互关联的过程组成的，如果仅关注任何单一的组成部分，对其理解都将是支离破碎的。因此，我们需要从更宽泛的角度来理解遥感过程以便于发现遥感技术所需要的知识(图1.1)。

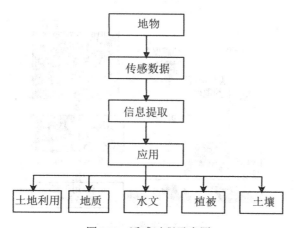

图1.1 遥感过程示意图

物理实体主要包括建筑物、植被、土壤以及水体等类似地物。当然也包含遥感应用人员所关注的其他地物。对地物的认知要视不同学科而定，比如地质学、森林学、土壤科学、地理学以及城市规划等。

传感数据是传感器如相机或雷达记录观测目标的反射或发射信号得到。因为传感器探测的视角是俯视的、且其分辨率与人眼所见存在差异,同时传感器的探测波段很多位于可见光之外,因此对于多数人来说,传感数据比较抽象和陌生。因此,要使用传感数据,首先需要将传感数据解译为信息(即信息提取),方能用于解决实际问题,比如填埋场选址、寻找矿产资源等。这些解译结果即为从影像数据中提取的信息,提取过程主要为将传感数据转换为某一特定信息的数据。实际上,同样的传感数据,如果从不同的视角出发可产生不同的解译结果。因此,同一景影像既可以提供土壤的信息,同时也可以用于土地利用、地质等方面,这要视特定的影像及不同的分析目的而定。一般而言,遥感所能获取的信息主要包括:

(1)平面位置和空间展布。
(2)地貌(高程)位置。
(3)颜色(光谱反射)。
(4)地表温度。
(5)纹理。
(6)表面粗糙度。
(7)湿度信息。
(8)植被生物量。

最后,我们来看一下应用。应用遥感数据时可以与其他形式的数据融合在一起来分析特定的实际问题,比如土地利用规划、矿产勘探以及水质测图等。

1.3.2 遥感过程的实例探讨

下面来看一下遥感影像生成的过程(图1.2)。

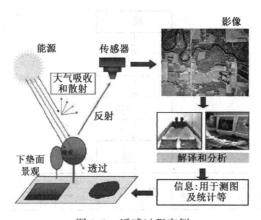

图1.2 遥感过程实例

(1)入射的太阳辐射部分地被大气所散射和吸收。
(2)剩余的辐射到达地表,与地物(本例中为一棵树)相互作用。
(3)大部分树冠的能量反射受控于单个叶片与辐射的相互作用,主要包括选择性吸收、透过和反射能量等,这些作用是与波段相关的。

（4）反射能量在达到传感器之前再次遭到大气的衰减。
（5）这种作用强度最终以类似图片的影像或者是由定量数据组成的矩阵呈现出来。
（6）影像经过解译和分析后，就可以获取地表相关的信息。

1.4 遥感的关键概念

遥感这一学科还很年轻，很多基本的事实和方法还没有完全地被人们所理解。科学家仍然在努力研究许多与遥感相关的基本方法和核心概念。尽管如此，对于我们学习遥感来说，还需要基于一系列的基本原理来理解遥感及其应用的本质问题。

1.4.1 空间分辨率

对于很多传感器而言，传感平台与遥感目标之间的距离对于获取目标区域的影像及目标的详细信息至关重要。传感平台上的传感器距离目标很远，一般视域很大，但目标物不够详细。将宇航员在航天器上所看到的与坐在飞机上的人所看到的进行对比，宇航员一眼就可能看到一个省份或者国家，但是却分辨不出单个的房屋。而当你乘坐飞机飞过一个城市或城镇时，你可以分辨出单个的建筑或者车辆，但是你所看到的区域却远远小于宇航员所观察到的区域。

影像的详尽程度依赖于传感器的空间分辨率。空间分辨率指的是可以分辨出的最小特征的尺寸。被动传感器的空间分辨率主要取决于他们的瞬时视场角（IFOV）（图1.3）。IFOV指的是传感器视域的立体角，它决定了传感器某一时刻在某一高度处所能探测到的地表的区域。地表的这一区域即为分辨单元，它决定了传感器的最大空间分辨率。

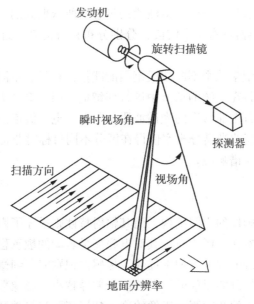

图1.3 垂直扫描仪示意图

粗（或低）分辨率的遥感影像只能分辨出大的目标物，而高分辨率的影像则可以分辨出小的目标物。以军用传感器为例，它需要探查的地物越详细越好，因此需要非常高的空间分辨率。而商用的卫星可以提供从几米到几公里空间分辨率范围内的影像。一般而言，分辨率越高，地面可视区域越小。

尺度指的是影像或地图上的距离与地面上所对应的实际距离的比值。如果一幅地图，其比例尺为1：100000，那么一个在地图上长1 cm的地物，其在地面上的实际长度为100000 cm（1 km）。地图与地面上长度的比值，其数值较小的影像或地图称为小尺度（比如1：100000）地图或影像，而具有大比值的（比如1：5000）则称为大尺度地图或影像。

空间分辨率也可以描述为卫星影像一个像元所代表的地面范围的大小。比如，Landsat卫星所搭载的专题制图仪（TM）的空间分辨率为30 m。一般而言，气象卫星传感器的地面分辨率通常大于1 km^2。

目前也有卫星的空间分辨率高于1 m，但是它们多为军用卫星，或者是价格昂贵的商业卫星系统。

1.4.2 光谱分辨率

影像上不同的地物类别或者其详细程度通常可以通过比较它们特定的波段响应范围来区分。大的类别，比如水体或者植被，一般可以通过比较宽的波段范围即可见光和近红外就可以区分开来。而其他更详细的类别，比如，不同的岩石类型，通过上述宽波段一般很难区分，需要更详细的波段范围才可以区分它们，因此我们需要高光谱分辨率的传感器。

光谱分辨率描述了传感器所能分辨的波段间隔的能力。光谱分辨率越细，某个通道或波段的波长范围越窄。

最简单的光谱分辨率的传感器仅有一个可见光波段。其影像与从飞机上所拍摄的黑白照片类似。在可见光波段具有三个光谱波段的传感器所获取到的信息与人眼视觉系统所获取的信息类似。TM传感器具有7个波段，分别分布在可见光、近红外、中红外和热红外光谱范围。

很多遥感系统可以记录几个不同波段范围的能量，且具有不同的光谱分辨率。这些传感器为多光谱传感器，将在后续的章节中进行详细地介绍。更高级的多光谱传感器为高光谱传感器，可以探测上百个非常窄的光谱波段，分布于电磁波谱的可见光、近红外和中红外部分（图1.4）。这样高的光谱分辨率使得在区分不同目标时非常有利，因为其可以探测目标物在每个窄波段的光谱响应。

1.4.3 辐射分辨率

像元的排列描述了影像的空间结构，而其辐射特性则描述了影像所包含的地物辐射信息。胶卷或者传感器每次获取影像时，其对电磁能量震动的敏感程度决定了传感器的辐射分辨率的大小。成像系统的辐射分辨率描述了传感器识别能量细微差异的能力。传感器的辐射分辨率越高，它所探测的反射或辐射的能量差异越小。影像数据通常使用正整数来表征，其变化一般从0到2的几次幂，影像的这一数据范围与二进制编码的位数相对应。每一位记录2的一次幂（比如1 bit=2^1=2）。因此，如果一个传感器用8位来记录数据，则

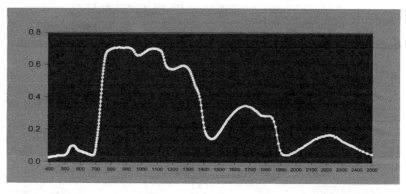

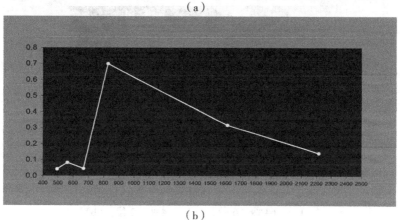

图 1.4 高光谱分辨率(a)和低光谱分辨率(b)

为 $2^8=256$,数字范围从 0 到 255。然而,如果用 4 位来记录,则 $2^4=16$,数值从 0 到 15。因此,相对于 8 位影像,其辐射分辨率就小得多。

通过对比一个 2 位和 6 位的影像,可以看出二者在分辨地物的细微程度上存在显著差异,这与影像的辐射分辨率密切相关(图 1.5)。

辐射分辨率表征了卫星能够分辨其所接收到的辐射的精细程度。通常采用位数来存储最大的辐射量,比如 8 bits 代表 256 个数值(通常为 0~255)。

- 1 bit （0~1）
- 8 bits （0~255）:每个像元都被拉伸到 0~255,0 表示没有光谱或者低于最小阈值,255 表示最大光谱响应或者高于 8 位数据的阈值。在 8 位影像上,通常表现不出地物的细微差别
- 16 bits (0~65535):具有更大范围的选择,但是其存储空间却是 8 位影像的 2 倍
- 32 bits (0~4294967295)及更多

图1.5 64级(6位,(a)),4级(2位,(b))

1.4.4 时间分辨率

除了空间、光谱和辐射分辨率,时间分辨率的概念也很重要。卫星传感器的重访周期(即时间分辨率)通常为几天。

时间分辨率指的是连续两次在同一地区进行数据获取的时间间隔。比如,在1971年、1981年、1991年和2001年分别进行一次航拍,则其时间分辨率即为10年。

时间分辨率表征的是一颗卫星访问同一地方的频率。其取决于:

(1)轨道特性。
(2)扫描带宽度。
(3)定位扫描设备的能力。

对于气象卫星而言,时间分辨率一般为一天几次,对于中等分辨率的卫星来说,时间分辨率一般为一年8~20次,如Landsat卫星。时间分辨率主要取决于卫星传感器的设计及其轨道运行模式。

1.4.5 几何畸变

每个遥感影像表征了一个特定几何关系下的景观,这一几何关系是由遥感传感器的结构、特定的运行环境、地表起伏以及其他的因素构成的。理想的传感器所记录的影像应该与地表保持准确的、固定的几何关系，这样的影像也是进行地表面积和距离量测的基础。当然，在实际应用中，每景影像都包含位置上的偏差，这主要是由于传感器探测的视角、扫描镜的运动、地形起伏以及地球曲率等造成的。每类错误在不同的情况下变化很大，只是，我们需要明白的是，几何上的错误是遥感影像所固有的特性，并非偶然误差。在某些情况下，我们可以去除或者降低位置上的误差，但是将影像用于进行面积或者距离等量测时，首先需要考虑几何畸变的影响。

1.4.6 图片和数字格式并存

多数遥感系统生成的地表影像为类图片格式。这样的影像通过系统地将其细分为相同尺度和形状的单元而形成数字格式，每个单元利用离散的数值代表该单元范围内的亮度值。相反的，有些遥感系统最初生成的就是数字矩阵，矩阵中的数值是以数字形式表达地表该区域的亮度值。数字影像通过将每个亮度值拉伸到一定数值范围内以图片影像的形式呈现出来。

图片和数字这两类遥感数据格式是利用不同的方法来显示和表达地物，但是它们之间在所表达的信息上没有实质的差别。根据应用目的的不同，可以选用不同的影像格式（有时，两种格式在相互转换时可能会损失一些细节）。

1.4.7 遥感系统

影像分析人员必须意识到的一点是：遥感过程的各个部分组成一个系统，因此，任何一个部分都不可能独立存在。比如，单一的升级相机透镜的质量意义不大，除非我们在升级相机透镜的同时也升级胶卷的质量，这样，分析人员从影像中提取的信息才会有所改善。

系统的各个组成部分应该与手头任务的需求相一致。这表明，解译人员不仅要确切掌握遥感系统，同时也要明确解译的目的、对细节的需求、数据获取的时间、最好的光谱区间等。就像系统的物理构成一样，解译人员的知识和经验也是相互作用并形成一个整体。

1.4.8 大气的作用

所有到达传感器的能量都需要穿过大气。对于可见光和近红外卫星遥感来说，传感器所接收的能量一定要经过相当深的地球大气。同时，太阳的能量在强度和波长上会随着大气中气体分子和颗粒的不同而发生变化，这些变化可能会造成影像质量的下降，进而影响解译的精度。

1.4.9 被动遥感和主动遥感

到目前为止，我们很多时候都把太阳作为能量或者辐射的来源。太阳为遥感提供了非常方便的能量来源，太阳的能量，在可见光部分被反射，在热红外部分，先吸收能量，然后发射出去。像这样依靠自然资源作为能量来源的遥感系统为被动传感器（图1.6(a)）。被动传感器只在自然能源可用时方能探测地表。对于所有的反射能量，只有当传感器处于太阳光照的一侧时方能被探测到。因此,在夜间,不能探测反射能量。那些无论白天还是黑夜发射的能量,比如热红外波段,只要其能量达到可以记录的量级,传感器就可以探测到。

与被动传感器形成对比的是主动传感器,它们自己提供自己的能量来源(图1.6(b))。此类传感器直接发射辐射到目标物,然后探测从目标反射回来的辐射。此类传感器的优势在于探测的全天候能力,与时刻、季节变化无关。主动传感器可以探测太阳不足以提供的波段范围,比如微波,或者可以控制照射目标物的方向。然而,主动传感系统需要能够产生足够多的能量以充分的照射地物。激光传感器和合成孔径雷达即为主动传感系统。

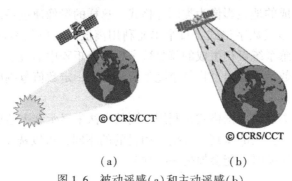

(a)　　　　　　(b)
图1.6　被动遥感(a)和主动遥感(b)

1.5　遥感的应用

遥感数据在实际应用中,一般需要与其他类型的数据关联起来,比如地形数据,行政边界和土壤、地质或者水文数据等。近些年,这些关联越来越趋向于在GIS框架下执行。尽管在GIS和遥感之间没有一个明确的界限,但是认为遥感主要用于采集数据,而GIS则主要用于存储和分析数据是片面的。

在地方级别的探测中,遥感影像记录了大尺度的地形、排水系统以及公路、建筑物、附属设施等基础设施。在国家和区域级别的探测中,遥感数据为制定宏观的发展模式,协调交通用地、居住用地、工业用地以及娱乐用地之间的关系,确定填埋场的选址,制定远景规划等提供依据。洲际间的遥感数据可以为宏观的自然资源勘查、环境问题监测包括土地开垦和水质变化,以及经济发展规划等提供信息。

在更大范围的探测中,各国政府利用遥感和影像分析的方法进行国内外环境的监测、土地的管理、农作物估产、救灾,以及为国家安全和建立国际关系等提供支持。遥感也用于支持国际间的活动,包括宏观尺度环境问题的分析、国际间合作与发展、救灾、为难民提供帮助,以及进行全球尺度的环境问题调查等。

在各个学科领域,遥感影像几乎与所有其他格式的数据相结合。地质学家和地球物理学家利用遥感影像研究岩相、结构、地表过程以及地质灾害。水文学家利用遥感影像识别土地覆盖模式、土壤水分状况、排水系统、河湖沉积物容量、海流以及水体的其他特征。

地理学家和规划人员利用影像研究聚落形态、土地资源调查、人类景观变化踪迹等。森林学家利用遥感和 GIS 进行木材的测图、估算木材的体积、监测病虫害、扑灭森林火灾，以及规划木材产量等。农业学家利用遥感影像进行农作物长势、成熟期和收割期的监测，同时监测灾害、病虫害、干旱的发生发展，以便于预测这些灾害对作物产量的影响。土壤学家利用遥感影像进行土壤类型成图，监测土壤模式与土地利用和植被之间的关系。简言之，遥感影像已经应用于各个领域来获取地表信息以用于自然或人类资源分布的分析。

☞ 主要知识点

- 遥感的定义
- 遥感传感器的类别
- 空间分辨率、光谱分辨率、辐射分辨率和时间分辨率、主动和被动遥感

☞ 思考题

（1）遥感数据的获取均是高空扫描得到的。课堂上已经对其优点进行了讨论，那么你能列举遥感影像的一些缺点吗？

（2）遥感影像是很多地物要素的综合，包含植被、地形、亮度、土壤等信息。那么，遥感影像的这种信息的综合是优点还是缺点呢？请给出原因。

（3）仔细思考空间分辨率、光谱分辨率和辐射分辨率三者之间的相关关系。是否可以提高三者中的一个而不会影响其他另外两个呢？

第 2 章 电磁波谱

任何物体只要其温度超过绝对零度，就会发射电磁辐射。同时，物体也会反射其他物体发射的辐射。通过记录发射和反射辐射能量，并考虑其穿过大气时与大气的相互作用，遥感分析人员获得了植被、建筑物、土壤、岩石以及水体等地表特征的知识(图 2.1)。

遥感影像解译需要理解电磁辐射及其与大气和地表的相互作用。

理解电磁辐射理论，为本课程后续内容的学习奠定了基础。

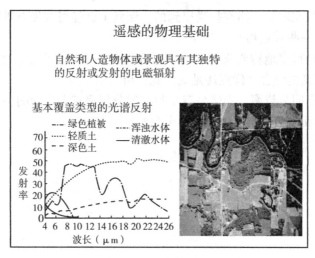

图 2.1 几种基本地表覆盖类型的反射光谱

2.1 电磁波

电磁场是由电场(E)和磁场(H)组成的。电场和磁场彼此正交，沿着垂直于传播方向的轴行进，并时刻发生变化(图 2.2)。

2.1.1 电磁波的三个属性

波长：波长是两个相邻波峰或波谷之间的距离。尽管波长可以利用日常的长度单位进行量测，但是因为在短波部分，两个波峰之间的距离即波长很小，因此对其测量采用极短的测量单位(如 nm，Å)。

频率：频率指的是在给定时间内通过某一特定点的波峰的数量。单位为赫兹。1Hz 即 1 圈/秒。

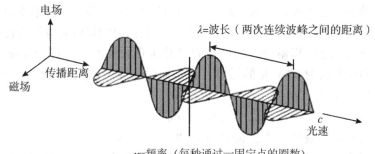

图 2.2 电磁波

振幅：振幅为电磁波每个峰值的高度。通常用能量水平（比较正式的为光谱入射辐射）来量测。

图 2.3 展示了振幅、频率和波长。中间的图表征频率高、波长短，最下面的图表征频率低、波长长。

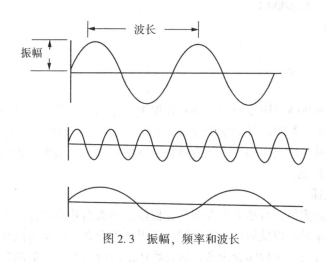

图 2.3 振幅，频率和波长

电磁能量的速度（c）为一恒定值，即 299893 km/s。频率（v）和波长（λ）关系如下：

$$c = \lambda v$$

频率与波长成反比，这表明波长越长，频率越低；反之，波长越短，频率越高。

波长和频率都可以用来描述电磁能量的特性。不同的学科，不同的应用有不同描述电磁辐射的传统，要么用波长，要么用频率。

因为没有一个权威的标准，所以光谱区域划分一直是遥感领域讨论最多的一个问题。

2.1.2 辐射定律

1. 斯忒藩-玻尔兹曼定律

太阳以及所有的地表物体如水体、土壤、岩石和植被，只要其温度超过绝对零度

(0 K)都会发射电磁辐射。太阳是电磁辐射的首要能源,也是很多遥感系统探测的主要部分。雷达和声呐系统是例外,因为它们发射电磁能量。辐射能量的大小及波长取决于物体的温度,物体温度升高时,其辐射的能量增强,最大辐射能量所对应的波长变短。这些关系可以通过黑体的概念予以解释。黑体指的是对所有波长的辐射都全部吸收和发射的物体。比如,太阳是温度为 6000 K 的黑体。黑体辐射的能量(单位为 W/m²)与其绝对温度的四次方成正比:即

$$M_\lambda = \sigma T^4$$

其中,σ 是斯忒藩-玻尔兹曼常数 5.6697×10^{-8} W·m^{-2}·K^{-4}.

这一定律阐述了地球(或太阳)发射的能量是其温度的函数。温度越高,其辐射的能量越大。

2. 维恩位移定律

辐射的最大波长可以通过下式进行计算:

$$\lambda_{max} \cdot T = k$$

其中,k 为常数 $= 2898$;T 为绝对温度,单位为 K.

比如,我们要计算太阳在 6000 K 和地球在 300 K 时,辐射的最大波长:

$\lambda_{max} = 2898$ μm·K/6000 K

$\lambda_{max} = 0.483$ μm

&

$\lambda_{max} = 2898$ μm·K/300K

$\lambda_{max} = 9.66$ μm

图 2.4 为具有 6000 K 温度的太阳、300 K 温度的地球以及其他几种地物的黑体辐射曲线。曲线所包围的区域为地物所辐射的总能量(M_t)。由于太阳的温度比地球高很多,因此太阳辐射的能量较地球要多很多。随着地物温度的升高,其最大光谱辐射所对应的波长(λ_{max})向短波方向移动。

3. 基尔霍夫定律

黑体是一假想的实体,自然界中并不存在黑体。因为自然界中任何实体都仅仅反射一部分入射辐射,并非是理想反射体。虽然理想的黑体不存在,但是可以在实验室条件下通过实验设备观测其行为。这种实验设备是研究物体的温度与其所发射的辐射量之间关系的基础。

基尔霍夫定律阐述了所有黑体在同一温度下其发射辐射与吸收的辐射通量的比是相同的。这一定律为定义发射率(ε)奠定了基础。发射率定义为一给定物体其发射辐射(M)与同一温度下黑体辐射(M_b)的比值,如下:

$$\varepsilon = M/M_b$$

黑体的发射率为 1,而理想反射体(白体)的发射率则为 0。黑体和白体都只是概念上的,但是可以在实验室人为的近似构造出来。在自然界中,任何实体的发射率均介于黑体和白体之间,即自然界的实体都是灰体。发射率这一参数可以用来测量灰体反射电磁能量的高低。对于灰体而言,其吸收入射辐射能量越多,其发射辐射也就越多,即具有较高的发射率。同样,吸收越少,发射越少,发射率也就越低,即这样的灰体发射能力强。

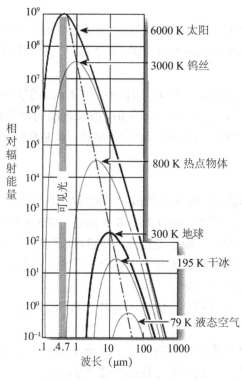

图 2.4 黑体辐射曲线

2.2 适合遥感的波谱范围

电磁波谱是电磁辐射连续变化的范围,从伽马射线(频率最高、波长最短)到无线电波(频率最低,波长最长),当然也包含可见光。

电磁波谱可以划分为 7 个不同的区域,分别是——伽马射线,X 射线,紫外,可见光,近红外,红外,微波和无线电波,如图 2.5 所示。

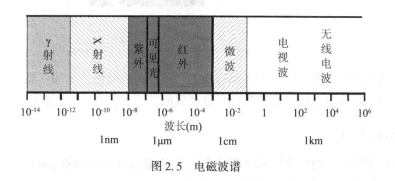

图 2.5 电磁波谱

2.2.1 紫外 (UV)

紫外区域波长范围 0.03~0.4 μm。这是用于遥感探测的最短的波段。

紫外可进一步详细划分为远紫外 (0.01~0.2 μm)，中紫外 (0.2~0.3 μm) 和近紫外 (0.3~0.4 μm)。

太阳是紫外辐射的自然能源。然而，波长小于 0.3 μm 的紫外辐射因为大气层中臭氧层的强烈吸收而不能达到地表。仅有 0.3~0.4 μm 波长范围的紫外辐射部分可以用于陆地遥感。这个光谱波段对某些物质所含的叶绿素荧光敏感，在某些特殊领域具有重要意义。

2.2.2 可见光(V)

电磁波谱中的可见光波段(0.4~0.7 μm)是遥感中使用最普遍的波段范围，也是我们人类肉眼可见的波段范围。其中，蓝 (0.4~0.5 μm)，绿 (0.5~0.6 μm) 和红 (0.6~0.7 μm)分别代表了加色法三原色，即通过这三种颜色可以混合成其他任何一种颜色。

尽管太阳光看似均一的颜色，实际上它是由不同波长的辐射组成的，主要是位于电磁波谱的紫外、可见光和近红外部分。

我们所看到的颜色大部分都是组成白色的各个不同波长范围的光经过反射和吸收后的结果。

比如，健康植被的叶绿素选择性地吸收了蓝光波段和红光波段以用于光合作用，反射了大部分的绿光波段，因此，我们所见到的植被多呈绿色。

雪之所以呈现白色，是因为雪将各种波长的入射光都散射掉的结果。

图 2.6 显示了地物上反射的加色法三原色即蓝色、绿色和红色。三原色不能由任何其他颜色混合而成。

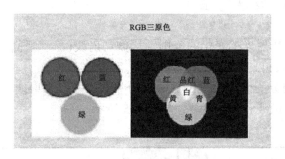

图 2.6　加色法三原色

三原色经过物体反射后进行组合就形成了各种不同的颜色。比如，品红是由红色和蓝色混合而成的，浅蓝色是由蓝色和绿色混合而成的，而黄色是由红色和绿色混合而成的。

2.2.3 红外(IR)

红外是位于可见光的红波段和微波之间的波长范围(0.7~1000 μm)。"红外"，从字面上理解，即为"红之外"，因为其与红光毗邻。

反射近红外(NIR)，波长范围为 0.7~1.3 μm，该范围的反射光可以用于黑白红外和彩色红外感光胶卷。

中红外(MIR)波长范围为 1.3~3 μm。中红外的能量可以利用光电传感器来进行探测。

热红外或者远红外(TIR)，波长范围 3~1000 μm。然而，由于大气的衰减，仅有 3~5 μm 和 8~14 μm 两个波段范围可以用于遥感。

热红外区域与热辐射息息相关。热能是由大气和地表所有地物持续热辐射的结果。光机扫描仪和特殊的摄像系统可以用于记录电磁波谱范围内的这部分热辐射。

2.2.4 微波(MV)

微波波长范围为 1 mm~1 m。微波可以穿透云、雨、植被冠层和干燥的表面沉积物等。

在微波区域有两类主要的传感器即被动微波传感器和主动微波传感器。被动微波系统探测地表辐射的自然微波辐射。主动微波遥感系统即雷达(Radar，即无线电探测和测距)人工发射微波辐射，并记录经由地表反射的这部分发射辐射。

图 2.7 为遥感使用的波段示意图。

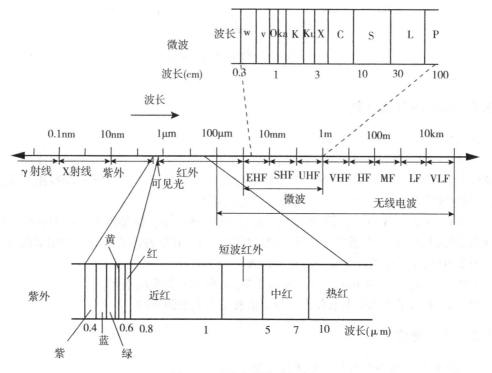

图 2.7 遥感中所使用的波段

2.3 典型地物的光谱曲线

图 2.8 为几种物质的光谱曲线。

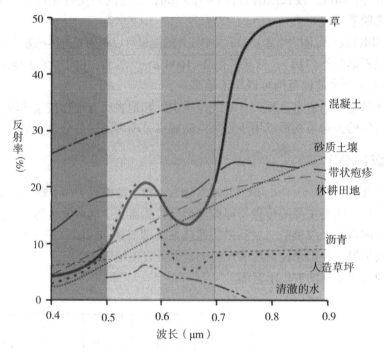

图 2.8 几种物质的光谱曲线

2.3.1 植被的光谱曲线

图 2.9 显示了植被的光谱曲线。

可见光（0.4~0.7 μm）：低反射率、低透过率，高吸收，这是由于在蓝波段中心区和红波段中心区叶绿素的存在所致。由于其他叶色素如叶黄素、类胡萝卜素和花青素的影响，在黄-绿波段中心区（0.55 μm）存在一个小反射峰。

近红外（0.7~1.3 μm）：低吸收率、高反射率和透射率，这是由于叶色素和细胞壁的纤维素是透明的缘故。在近红外区域形成高反射区域，在红波段的边缘具有明显的反射凸起。图 2.10 展示了植被红边效应所在的位置。

中红外部分，在 1.4 μm、1.9 μm 和 2.7 μm 处对水的吸收很强。

图 2.11 展示了玉米叶中水分的含量对植被反射光谱的影响。

2.3.2 水的光谱

水的光谱与水的本性及其所处状况有关。

遥感中水体的识别更多的是在近红外波段中实现的。

在自然条件下，在近红外和中红外波段，水体吸收了所有的入射光。

在可见光部分，水体的物质能量交换是十分复杂的，比如反射可能来源于水体的表面、水体的沉积物或者是水体中的悬浮物。

图 2.12 显示了浑浊水体和清澈水体的光谱。

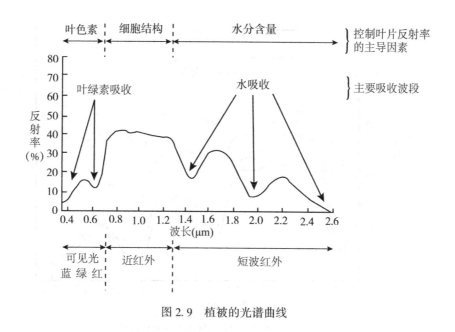

图 2.9 植被的光谱曲线

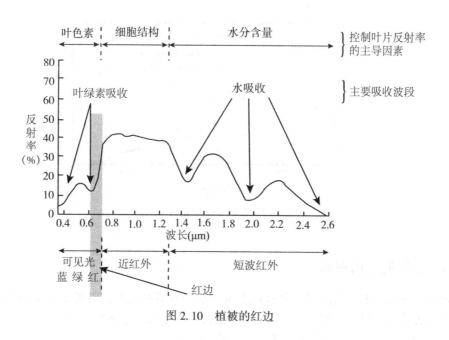

图 2.10 植被的红边

2.3.3 雪的光谱

在 0.5~1.1 μm 之间，由于雪和云光谱的相似性，很难将二者区分开来。因此，测定雪的范围及其状况最好是利用中红外波段。图 2.13 为雪在不同时期的光谱曲线。

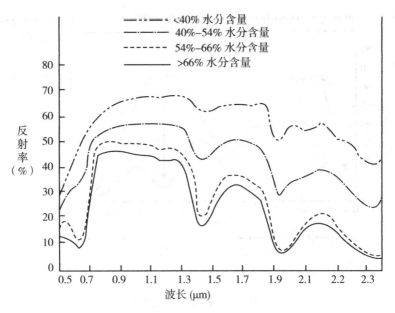

图 2.11 玉米叶中水分含量对其反射光谱的影响

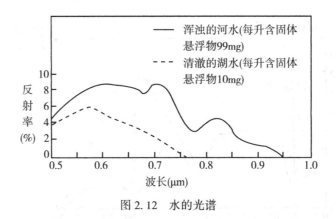

图 2.12 水的光谱

2.3.4 土壤的光谱曲线

影响土壤光谱曲线的因素主要有：土壤中水分含量的大小，有机质含量的多少，矿物或化学组成的多少以及土壤的粒径和纹理。图 2.14 为几种土壤的光谱。

图 2.15 为土壤水分含量对土壤光谱曲线的影响。

2.3.5 获取光谱的途径

- 利用光谱仪在野外实地测定光谱
- 在实验室测定
- 从影像数据中采集

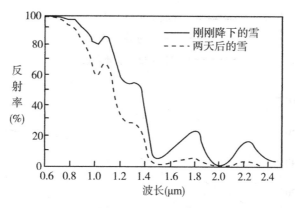

图 2.13 雪的光谱

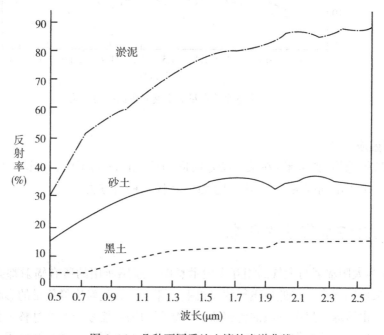

图 2.14 几种不同质地土壤的光谱曲线

- 光谱库网站中获取：http://speclab.cr.usgs.gov/spectral-lib.html

2.3.6 大气对光谱响应模式的影响

1. 时相效应

光谱特性随着时间变化而变化。比如：在植物整个生长过程中其光谱的变化。这些变化通常会影响我们为了实现某一特定应用而采集影像数据的时间。

21

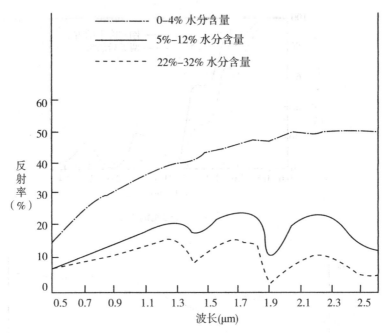

图 2.15 土壤水分含量对土壤光谱曲线的影响

2. 空间效应

同一种特征地物(如玉米)在某一特定时间随地理区域的不同,光谱亦发生变化。这种变化主要是由于不同区域具有不同的土壤、气候和种植方法。

2.4 大气对传感信号的影响

图 2.16 为太阳辐射与大气的相互作用示意图。遥感所用的所有辐射都必须经过地球大气层。如果传感器搭载在低空飞行的飞机上,那么,大气对影像质量的影响则可以忽略不计。相反,由地球卫星所搭载的传感器所采集到的辐射能量则要经过整个大气层。在这种条件下,大气对传感数据及影像质量的影响则很严重。因此,学习及使用遥感技术应该理解大气与电磁能量的相互作用过程。

图 2.17 所示为辐射能量到达传感器之前与大气相互作用的几种情况。

Ray-1:光子离开地表后没有发生任何改变而直接抵达传感器。这些辐射是对遥感有用的信号。

Ray-2:光子离开地表向传感方向辐射,但是由于被大气吸收而未能抵达传感器。

Ray-3:光子由于大气的散射而偏离在传感器的视场角范围之外。

Ray-4:由大气自身辐射出的光子抵达传感器。

Ray-5:来自主动或被动光源的光子没有接触地表而直接被大气散射到传感器的视场角范围内。

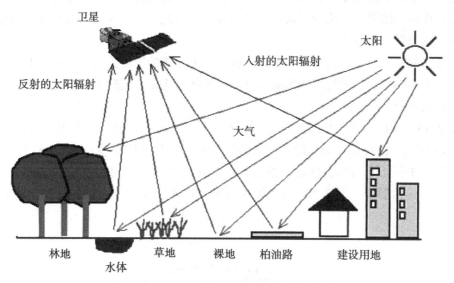

图 2.16 太阳辐射与大气的相互作用示意图

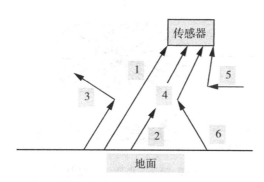

图 2.17 辐射与大气的相互作用过程

Ray-6：光子离开地表并携带有传感器视场角范围之外的信息，由于大气折射而散射入传感器的视场角范围之内。

测量大气的透过率，可以通过第 1 部分辐射携带的能量与第 1，2，3 部分辐射所携带总能量的比值来实现。如果第一部分辐射所携带的能量与第 4，5，6 部分所携带的能量相比少很多，那么遥感的可用性就大大降低。当然，可以经过适当的大气校正算法来实现对地表信号的增强。这类大气校正算法不但要去除 4，5，6 部分的贡献，同时还要估算 2，3 部分辐射的能量大小，以便于揭示地表真实的信号。

可见，当太阳辐射经过大气层时被几种物理过程所改变，主要包括散射、吸收和折射等。

2.4.1 散射

散射是大气中悬浮的粒子或大气气体中的大分子改变了电磁能量的前进方向。散射的发生与大气中粒子的大小、其富集程度、入射光的波长以及能量穿过大气的深度有关。

散射作用的结果就是改变了入射辐射的方向,因此一部分入射光束被返回到空中,同时也有一部分射向地球表面。

电磁辐射一旦产生,它在大气中以接近真空中光速的速度传播。与真空中光速传播不受任何物质影响相反,大气不但影响入射辐射传播的速度,同时也影响其波长、强度、光谱分布以及方向。

图 2.18 显示了三类大气粒子的散射行为。其中,图 2.18(a)表明大气烟尘形成较大的不规则粒子,其具有明显的前向散射峰,而后向散射则很少;图 2.18(b)表明大气分子在形状中接近于对称,因此其散射特性是前向和后向散射近乎相同,但是没有(a)中那么明显的散射峰;图 2.18(c)表明大的水滴产生明显的前向散射峰和较小的后向散射峰。

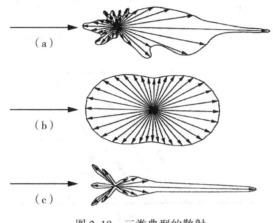

图 2.18 三类典型的散射

图 2.19 显示了三种类型的散射:瑞利散射、米氏散射和非选择性散射。

散射类型的发生主要受限于:(1)入射太阳辐射的波长;(2)辐射所经过的气体分子、烟尘颗粒、水蒸气/水滴的尺寸。

1. 瑞利散射

晴空大气中仅仅包含大气气体分子,使得光线的散射随着入射光波长的缩短而显著增强。瑞利散射是大气气体分子最易发生的散射行为。当入射光线的波长(λ)远远大于空气中的粒子(通常是气体分子)的直径(α)(即 $\alpha \ll \lambda$)时易于发生瑞利散射。其散射强度几乎与入射光波长的四次方的倒数成正比,图 2.20 为瑞利散射的散射强度与入射光波长的关系。瑞利散射也是天空成蓝色的原因。蓝光(波长 400 nm)的散射强度是近红外光线(波长 800 nm)散射强度的 16 倍。瑞利散射是大气中主要的散射形式,其主要发生在 9~10 km 高空。

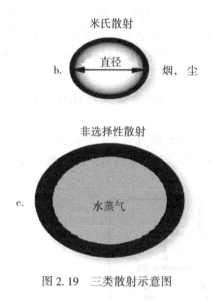
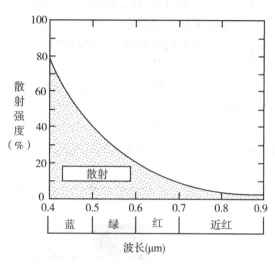

图 2.19 三类散射示意图　　　图 2.20 瑞利的散射强度与波长的关系

2. 米氏散射

米氏散射是由悬浮于空气中的大粒子包括尘埃、花粉、烟和水滴等引起的。这些粒子要比发生瑞利散射的粒子大得多。其粒子粒径的大小与入射光波长几乎相等。

米氏散射发生在从可见光及其附近波长范围内。米氏散射的散射强度与波长有关，但是其与波长的关系不似瑞利散射的散射强度与波长的关系那样简单。米氏散射在近地面范围内(0~5 km)发生，因为在这一范围，大粒径粒子比较容易富集。

3. 非选择性散射

当空气中粒子的粒径远大于入射辐射的波长(即 $\lambda \ll \alpha$)时，发生非选择性散射。在可见光及其附近区域的辐射，发生非选择性散射的粒子一般为大的水滴或者大粒径的飞机烟尘。

非选择性意味着非选择性散射与波长无关，因此，我们所见到的白色或灰色的阴霾天气就是由于所有的入射辐射被同等程度地散射的结果。

2.4.2 折射

折射是入射光线在两个介质的界面上穿过时发生弯曲的现象。此外，当大气由于其清洁度、浑浊度和温度等的不同而呈现分层时，光线穿过这样的分层大气也会发生折射。因为大气中各层的物质密度不同，反过来引起入射光线从某一层穿到另一层时也会发生弯曲。

典型的一个例子即为，在炎热的夏季，从远处看高速公路、跑道和停车场上空出现闪烁的现象，就是由于靠近这些硬地面的空气要较上空热很多，造成大气分层而出现折射引起的。

2.4.3 吸收

入射的太阳辐射被吸收并转换为其他形式能量的过程即为吸收。当入射辐射能量的频率与一个原子或分子的谐振频率相同的时候发生吸收，并使得原子或分子产生活跃状态。原子或分子并不是向外辐射具有相同波长的光子，而是将吸收的能量转换为热运动并以较长的波长向外辐射能量。

吸收波段分布在电磁波谱的范围很宽，通常辐射能量被各类物质如 H_2O，CO_2，O_2，O_3 和 N_2O 所吸收。

各部分的吸收累积效应可阻止某些光谱区域的入射辐射通过大气（图 2.21），这种现象对遥感很不利，因为在大气吸收强烈的区域，传感器感应不到任何有用的能量。

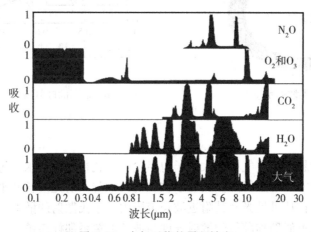

图 2.21 大气吸收的累积效应

太阳入射辐射能量可以被 0.1~30 μm 范围内的各类不同大气气体分子所吸收。臭氧在地球能量平衡中起着重要的作用。臭氧对具有高能量、短波的紫外光子的吸收可以阻止紫外线穿过低层大气。因臭氧对紫外的强烈吸收而使得人类免受紫外线辐射，从而避免皮肤癌的发生。

二氧化碳在遥感中很重要，因为其在中红外和远红外区域对辐射具有吸收作用。二氧化碳在位于热红外 13~17.5 μm 区域的吸收作用最为强烈。这也是温室效应形成的原因。

水蒸气的富集程度随着时间和区域位置的不同变化很大。水蒸气的吸收是其他气体吸收总和的几倍。水蒸气具有强烈吸收作用的区域位于 5.5 μm 和 7.0 μm 间的几个波段以及 27.0 μm 以外的区域。

1. 大气窗口

那些相对比较容易透过大气的波长范围即为大气窗口（图 2.22）。

图中只是示意性的给出了几个重要的窗口。阴影区域代表大气对电磁辐射的吸收。

大气窗口对遥感的作用显而易见，它们定义了可以用于成像的这些电磁波段。而未落

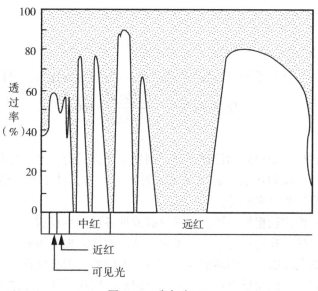

图 2.22 大气窗口

于大气窗口内的其他波段的能量,因为被大气严重衰减,因此对于遥感来说是无效的信号。

在远红外波段,两个重要的窗口范围是 3.5~4.1 μm,和 10.5~12.5 μm,后者尤其重要,因为地表发射峰的波长就位于这个波段范围内。表 2.1 为遥感可用的主要大气窗口。

表 2.1　　　　　　　　　　　遥感可用的主要大气窗口

紫外和可见光	0.30~0.75 μm;0.77~0.91 μm
近红外	1.55~1.75 μm;2.05~2.4 μm
热红外	8.0~9.2 μm;10.2~12.4 μm
微波	7.5~11.5mm;20.2+mm

2. 云

大多数电磁辐射均不能穿透云。云是遥感成像的一个大问题,在热带地区尤为显著。一般而言,多时相影像合成的方法可以去除云。此外,云的阴影也是一个问题。

2.5　大气辐射与地表的相互作用

当电磁辐射能量到达地表时,其要么被反射,要么被吸收,要么穿过地表,每个过程所占的比重主要取决于地表的自然特性、电磁能量的波长以及入射角度的大小。

2.5.1 反射率

1. 定义

反射率是在某一波长范围内被反射的辐射与入射辐射的比值，可以表达为：

$$反射率(\%) = \frac{反射辐射}{入射辐射} \times 100\%$$

2. 反射类型

与入射波长相比，如果表面比较光滑，则发生镜面反射。镜面反射几乎将所有的入射光以一个相同的角度反射出去。就可见光入射辐射而言，易于发生镜面反射的表面有镜子、金属表面或者是平静的水面。

相反，如果表面比较粗糙，则发生漫反射或称各项同性反射。这时，反射的能量几乎在各个方向是相等的。在可见光波段，很多自然表面都表现为镜面反射体，比如，相对均一的草地。理想的漫反射体，即朗伯体，在各个方向的反射亮度都是相等的。

多数地表既非完全镜面反射体，也非完全朗伯反射体。任何给定表面，描述其反射类别，主要依据相对于感应能量波长，该表面的粗糙度的大小。

漫反射包含了反射表面颜色的光谱信息，而镜面反射则通常不包含其表面的光谱信息。

在遥感中，我们更多的是测定地形特征的漫反射特性。

此外，反射率随着波长、几何关系变化而变化，并且是判断不同物质的依据。

2.5.2 透过率

当辐射穿过一个物质而没有发生明显衰减时，即为透过。给定一物质的厚度，能量穿过该物质的能力可以用透过率(t)来计算：

$$t = \frac{透过的辐射}{入射辐射}$$

在遥感领域，薄膜和滤镜的透过率比较关键。对于自然界的物质来说，我们一般认为辐射对水体的穿透能力较强。然而，许多物质的透过率随波长变化很大，因此，我们眼睛所见到的可见光谱才不会相互混淆。比如，可见光部分的辐射几乎不能穿过植被叶片，而红外辐射则比较容易穿过。

☞ **小结**

本节主要讲述了电磁波谱、斯忒潘-玻尔兹曼定律和维恩位移定律、光谱以及大气和地表对电磁辐射的作用。

☞ **主要知识点**

- 电磁波的三个特性
- 斯忒潘-玻尔兹曼定律

- 维恩位移定律
- 基尔霍夫定律
- 植被、水的光谱特性
- 散射
- 折射
- 反射
- 大气窗口

☞ **思考题**

(1) 请结合瑞利散射解释晴朗的天空呈蓝色以及日出时天空呈明亮的橙红色的原因。
(2) 结合本节课程内容，谈谈为什么有些路灯被设计成红色的。

第3章 遥感平台和传感器

3.1 简介

世界气象组织(WMO)全球观测系统(GOS)可以观测和收集全球范围内的天气、水和气候信息(如图3.1所示)。该系统由14颗卫星、上百个海洋浮标、飞机、轮船以及近10000个地面观测站组成。国家气象和水文服务(NMHSs)生产和收集各自国家的观测数据。负责联系全球各个国家的国家气象中心的WMO的全球电信系统(GTS),每天发布50000多个天气报告、几百个图表以及数字产品。

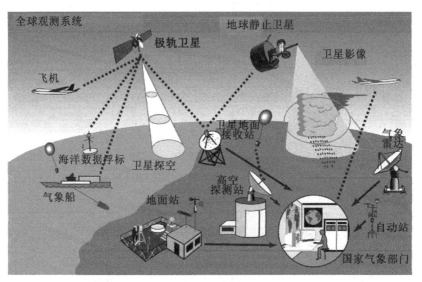

图3.1 全球观测系统

3.2 遥感平台

- 地面平台:其高度一般小于100 m。
- 航天飞机:按飞行高度可以分为低空、中空和高空。空间分辨率较高。
- 卫星:极轨太阳同步卫星的轨道高度为800~900 km,绕地一周需90~100分钟。地球同步卫星:相对地球静止,轨道高度为35900 km,绕地球一周需24小时。

3.2.1 轨道和扫描带

卫星所飞行的路径即为其轨道。在任何时候都观测地球同一位置的卫星为地球静止卫星。气象和通信卫星通常是地球同步卫星。许多卫星设计成为沿南北轨道飞行，因为地球一直自西向东运转，所以这类卫星在一定时间段内可以覆盖整个地球。这类卫星是近极轨卫星。它们大多是太阳同步的卫星，因此这类卫星通过某个地方的当地时间几乎是不变的。极轨卫星在地球的一侧自南向北运转时，称为上升轨；而在另一侧时则是自北向南运转，称为下降轨。

当一颗卫星绕地球旋转时，传感器可以探测到地球表面的某一部分，成像的这一部分即为扫描带。位于卫星正下方的地面被称为星下点。当卫星沿轨道运行时。具备立体观测的卫星可以在星下点观测之前和(或)之后对观测目标进行两次或三次观测。

3.2.2 太阳同步卫星

卫星轨道与太阳同步，和太阳始终保持相对固定的方向。

1. 极轨

极轨是一个特定的低轨类型，卫星沿南北向而非东西向运转。

2. 使用极轨的好处

极轨非常有助于探测星球的表面。卫星沿轨道南北向飞行，地球自西向东自转，这样极轨卫星就可以扫描到整个地球表面。

3. 轨道倾角

极轨轨道倾角为 90°，它与赤道面垂直。

4. 近地轨道

当一颗卫星沿近地球轨道运转时，这样的轨道称之近地轨道(或低地轨道)，简称低轨。低轨卫星一般 200～500 英里(320～800 km)高。因为这样的轨道距离地球很近，所以它们必须快速旋转，否则地球的重力就会把它们拉回到大气层。低轨卫星的轨道速度为每小时 17000 英里（27359 km），绕地球一圈需要约 90 分钟。

5. 视域广

低轨非常有用，因为它近地球运转，所以视域很壮观。遥感和气象卫星在观测地球表面时，经常低轨运行，因为在这个高度上他们能够获取地表的细节影像。

6. 太空垃圾

低轨太空环境目前已经非常拥挤。美国空间站(USSC)可以探测出轨道上卫星的数量。根据 USSC 的报道，低轨太空上存在有 8000 多个比垒球大的物体。

极轨具有较低的高度，影像的分辨率较高。

这些卫星在返回同一地点之前绕地球旋转了很多圈。

3.2.3 地球静止卫星

地球静止卫星能够从高空观测地球。因为地球静止卫星与地球转动速度相同，所以它们能对同一地区进行长期探测，这有助于进行预报预测。

地球静止卫星的轨道为平行于赤道平面的圆形轨道，卫星需要定位在一定的高度以保证卫星的轨道周期与地球旋转的速度相匹配，因此卫星看起来好像悬挂在地球赤道上空的一个亮斑。

虽然地球静止卫星是以赤道为中心的固定地理区域进行重复观测的理想卫星，但是它们距离地球很远，因此很难获取高质量的定量观测数据。现役的地球静止气象卫星确实是一个技术奇迹。但是，这些卫星却看不到极地地区，而且一般需要5~6颗卫星才能覆盖赤道一圈。

3.3 常用遥感数据源

3.3.1 低分辨率遥感数据

1. AVHRR（高级甚高分辨率辐射计）

最初的设想是用于气象，但目前已能够应用于很多领域。

2. 光谱分辨率

- 0.58~0.68 μm（红波段）
- 0.725~1.10 μm（近红外）
- 3.55~3.93 μm（热红外）
- 10.3~11.3 μm（热红外）
- 11.5~12.5 μm（热红外）

优点：快速的全球覆盖（高时间分辨率），数据处理工作较少。

缺点：空间分辨率较低。

3. AVHRR 的应用

(1)植被监测和荒漠化探测：主要基于植被指数或归一化植被指数实现植被和荒漠化的监测。其中，植被指数=波段2-波段1；归一化植被指数为(波段2-波段1)/(波段2+波段1)。

(2)全球变化研究：气候变化及监测。

3.3.2 中分辨率卫星

1. 美国的地球观测系统

地球观测系统(EOS)是地球科学事业的一部分，NASA的主要意图是评估自然事件和人类活动对地球环境的影响。包括太空和地面观测系统两部分。

EOS 的 TERRA 卫星上有 5 个主要的传感器：

ASTER——可见/近红外传感器，可立体成像，可生产 DEM。

CERES——辐射平衡测量（主要用于全球变暖）。

MISR——地球的多角度观测，获取大气数据。

MODIS——"新一代 AVHRR"，36 个通道（波段），空间分辨率分别为 250 m，500 m 和 1 km。

MOPITT——对流程污染测量仪。

2. MODIS

- 中分辨率成像光谱仪
- 搭载在 Terra 卫星上
- 可以获取 36 个光谱波段的数据，从 0.4~14.4 μm
- 空间分辨率：有 2 个波段为 250 m，5 个波段为 500 m，29 个波段为1 km
- 轨道高度 705 km
- 扫描带宽度 2330 km
- 1~2 天可完成全球覆盖一次
- 网址：http：//modis.gsfc.nasa.gov

3. ASTER

- 高级星载热辐射仪和反射仪
- 搭载在 Terra 卫星上
- 包含 3 个子系统：3 个可见光和近红外波段（VNIR），分辨率为 15 m。6 个短波红外波段（SWIR），分辨率为 30 m。5 个 热红外波段(TIR)，分辨率为 90 m

图 3.2 揭示了 Terra 卫星上搭载的传感器与其科学目的、观测数据之间的关系。

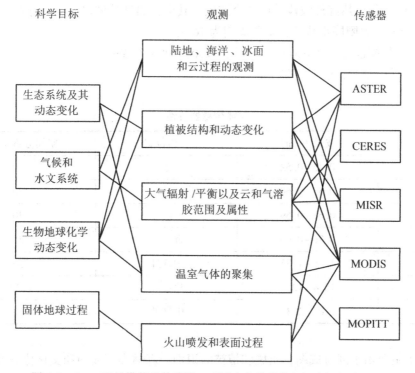

图 3.2 Terra 卫星搭载的传感器、观测内容及科学目标之间的关系图

4. Landsat 卫星

截至目前，已经发射了 7 颗 Landsat 卫星，其中，3 颗在轨运行（Landsat 4，Landsat 7，Landsat 8）。Landsat 6 因发射失败而未能正常工作。

- 由美国国家宇航局（NASA）代理
- 极轨卫星，轨道高 705 km，太阳同步
- 扫描带宽度 185 km
- 重访周期为 16 天

Landsats 1，2，3

地球资源技术卫星（ERTS-1），后来命名为 Landsat 1，于 1972 年发射升空，主要目的是为了测试卫星接收地球资源数据的可能性。

（1）所接收的数据为全球共享即实行 OpenSkies 策略服务。

（2）搭载了一台多光谱扫描仪（MSS），扫描宽度为 185 km，共 4 个波段，2 个可见光波段，2 个近红外波段。空间分辨率为 80 m，太阳同步，重复周期为 18 天。

Landsats 4，5

（1）Landsat 4 发射后不久就失效了，但仍然在轨。

（2）Landsat 5 搭载一颗多光谱扫面仪（MSS）和专题制图仪（TM），扫描带宽度为 185 km。

（3）在蓝波段至热红外波段共设 7 个波段。其中，热红外波段的空间分辨率为 120 m，其余均为 30 m，太阳同步轨道，重复周期为 16 天。

（4）每景数据包含 36000000 万个像元，250000000 个数据值。

表 3.1 所示为 TM 传感器的特性。

表 3.1　　　　　　　　　　　　　TM 传感器特性

波段	波长（μm）	光谱位置	空间分辨率(m)
1	0.45~0.52	蓝	30
2	0.53~0.60	绿	30
3	0.63~0.69	红	30
4	0.76~0.90	近红外	30
5	1.55~1.75	中红外	30
6	10.4~12.5	热红外	120
7	2.07~2.35	中红外	30

波段 1 主要用于滨海成图，土壤和植被的识别、森林类型成图和文化特征鉴定等。

波段 2 主要用于测量植被的绿反射峰，并将其用于植被区分、能量评估和文化特征鉴定等。

波段 3 主要通过探测叶绿素的吸收区域来区分植被物种、文化特征鉴定等。

波段 4 主要用于测定植被类型、能量含量和生物量含量，鉴定水体、土壤水分等。
波段 5 主要用于指示植被水分含量和土壤水分含量，区分云和雪。
波段 6 主要用于植被缺水分析、土壤水分含量测定和热测图等应用。
波段 7 主要用于矿物和岩石类型的区分，对植被水分含量很敏感。

那么，Landsat 系列影像的分辨率是如何逐年变化的呢？见表 3.2。

表 3.2　　　　　　　　　　　　Landsats 1~7 上的传感器

传感器	卫星	光谱范围(μm)	分辨率(m)
RBV	1, 2	0.475~0.575	80
		0.580~0.680	80
		0.690~0.830	80
	3	0.505~0.750	30
MSS	1-5	0.5~0.6	79/82[a]
		0.6~0.7	79/82[a]
		0.7~0.8	79/82[a]
		0.8~1.1	79/82[a]
	3	10.4~12.6[b]	240
TM	4, 5	0.45~0.52	30
		0.52~0.60	30
		0.63~0.69	30
		0.76~0.90	30
		1.55~1.75	30
		10.4~12.5	120
		2.08~2.35	30
ETM[c]	6	上述 TM 波段 plus 0.50~0.90	30(热红外波段 120m) 15
ETM+	7	上述 TM 波段 plus 0.50~0.90	30(热红外波段 60m) 15

注：[a] Landsats 1, 2, 3 的分辨率为 79m，Landsat 4, 5 为 82m。
　　[b] 发射后不久就失效了（Landsat 3 的第 8 波段）。
　　[c] Landsat 6 发射失败。

Landsat 7

(1) Landsat 7 于 1999 年 4 月发射升空。

(2) Landsat 7 携带有增强型专题制图仪(ETM+)，扫描带宽度为 185 km。

（3）全色波段分辨率为 15 m，其他可见光波段、近红外、中红外波段与 TM 相同，热红外波段空间分辨率上升为 60 m。

（4）同样是太阳同步轨道，重访周期为 16 天。

（5）EMT+传感器上的扫描线校正器（SLC）于 2003 年 5 月 31 日失效了，因此其后的 Landsat 7 影像大约丢失了 22% 的数据。

Landsat 8

（1）Landsat 8 于 2013 年 2 月发射升空。

（2）Landsat 8 携带有两个传感器：陆地成像仪（OLI）和热红外传感器。陆地成像仪（OLI）精细化 EMT$^+$ 的波段的同时，增加了三个波段：深蓝波段用于沿海气溶胶研究，短波红外波段用于卷云探测，以及一个质量评定波段。热红外传感器（TIRS）分为两个热红外波段，可以用于温度的精确计算。

（3）OLI 多光谱波段 1~7，9，空间分辨率为 30 m，波段空间分辨率为全色 15 m，热红外波段为 100 m。

（4）一景影像覆盖范围大约为 170 km（南北向）×183 km（东西向）。

（5）一景影像的压缩文件大小约为 1G。

5. SPOT 卫星

SPOTS 1，2，3

SPOT——SPOT 卫星由法国国家空间中心与其他欧洲组织所构想和设计，运营由法国负责。SPOT 1 于 1986 年 2 月 22 日发射升空，SPOT 2 于 1990 年 1 月 22 日升空，SPOT 3 于 1993 年 9 月 26 日升空，SPOT 4 于 1998 年 3 月 24 日升空，SPOT 5 于 2002 年 5 月 3 日成功发射。

SPOT 包含两个相同的传感设备，一个遥测发射机，一个磁带记录仪。这两个传感器称为 HRV（高分辨率可见光）传感器。HRV 有两种工作模式。全色和多光谱模式。在全色模式中，传感器的光谱响应波段范围为 0.51~0.73 μm，扫描带宽度为 60 km，一行包含 6000 个像元，空间分辨率为 10 m。在此模式中，HRV 传感器提供了较高的空间分辨率，但是其光谱范围较宽。在多光谱模式中，HRV 传感器的光谱区间为：

- 波段 1：0.50~0.59 mm（绿波段）
- 波段 2：0.61~0.68 mm（红波段；叶绿素吸收）
- 波段 3：0.79~0.89 mm（近红外波段；大气降水）

在这一模式中，扫描带宽度为 60 km，每行包含 3000 个像元，空间分辨率为 20 m。在此模式下，光谱分辨率很高，但空间分辨率很低。在某些情况下，如果位于同一区域，可以利用全色的高分辨率数据提升具有低分辨率多光谱数据的空间分辨率。

在扫描几何上，每个 HRV 传感器均可以进行天底观测和非天底观测。在星下点观测中，两个传感器均可以覆盖相邻的区域。因为两个 60 km 的扫描带有 3 km 是叠加的，因此，总的影像宽度为 117 km。在赤道地区，相邻卫星轨道中心最大相距 108 km，因此，在这一模式下，卫星可以实现对地表的全覆盖。

通过在地面对旋转镜发送指令可实现非天底观测。在非天底模式下，传感器可以观测

950 km 宽的范围，单景影像的扫描带宽度为 60~80 km，这主要取决于不同的观测角。反过来，相同的区域可以从不同的位置（即不同的卫星通过时刻）进行观测以获取立体覆盖。两个传感器不能同时以一种方式工作，即一个 HRV 为垂直模式，则另一个 HRV 为倾斜模式。SPOT 对同一地区重复观测的时间为 1~5 天，这要据不同纬度而定。

SPOTS 4，5

SPOT 4 和 SPOT 5 分别于 1998 年 3 月和 2002 年 5 月发射升空。SPOT 4 的主要特征是搭载有高分辨率可见光和红外（HRVIR）传感器，相对于 SPOT 1，2，3 所搭载的 HRV，HRVIR 增加了中红外波段（1.58~1.75mm），主要用于地质勘查、植被和雪覆盖的调查等。

SPOT 4 搭载了两个相同的 HRVIR 传感器，每个都可以相对地面轨迹两侧旋转 27°，可以获取 460 km 宽的重复地表覆盖或立体像对。在其单光谱模式（M）中，光谱波段 2 的空间分辨率为 10 m，多光谱模式（X）中，波段 1，2，3 和中红外波段的空间分辨率为 20 m。

SPOT 5 搭载了升级了的 HRVIR 传感器，其空间分辨率为 5 m，并可以获取沿路径立体像对。这是一新的具有高分辨率几何特性的传感器设备（HRG），也具有与 SPOT 4 相同的波段、空间分辨率和数据获取的能力，因此可以与早期的卫星组成持续的观测序列。

6. Landsat 和 SPOT 数据的应用

- 地质：矿产成图和石油勘探
- 农业：农作物估产，作物长势监测
- 林业：评估由火灾、过度砍伐和灾害造成的损失，林业资源调查及林业用地评估等
- 土地利用规划：现状图例覆盖成图、变化检测以及路线选址规划
- 替代高纬度航片
- 监测牧场生存状况、野生动物栖息地、鉴定水污染、鉴定洪泛区域、为自然灾害损失评估提供帮助等

3.3.3 高分辨率卫星

1. IKONOS

IKONOS 卫星系统，1999 年 9 月发射升空，由位于美国丹佛的空间成像有限公司运营管理。在全色模式下，IKONOS 的空间分辨率为 1 m，光谱范围为 0.45~0.90 μm。在多光谱模式下，其空间分辨率为 4 m，波段分别为：

波段 1：0.45~0.52 μm（蓝波段）

波段 2：0.52~0.60 μm（绿波段）

波段 3：0.63~0.69 μm（红波段）

波段 4：0.76~0.90 μm（近红外波段）

影像星下点的扫描带宽度为 11 km，太阳同步，通过赤道的时间是上午 10：30。重访周期随纬度不同而发生变化，在纬度 40°地区，多光谱模式重复覆盖时间大概为 3 天，全色模式为 1 天半。

2. Quickbird

美国的地球观测公司于 2001 年 10 月发射了 Quickbird 卫星，主要用于获取精细影像。1 个全色波段，空间分辨率为 0.61 m。4 个多光谱波段，空间分辨率为 2.44 m，具体如下：

波段 1：0.45~0.52 μm（蓝波段）
波段 2：0.52~0.60 μm（绿波段）
波段 3：0.63~0.69 μm（红波段）
波段 4：0.76~0.89 μm（近红外波段）
波段 5：0.76~0.89 μm（全色波段）

Quickbird 的扫描带宽度为 16.5 km。

3. GeoEye 1

GeoEye 1 于 2008 年 9 月 6 日发射升空，是目前世界上空间分辨率最高的商用地球成像卫星。

GeoEye 1 装备有其他商业卫星所没有的最复杂的技术，提供了前所未有的空间分辨率，全色波段 0.41 m，多光谱波段 1.65 m。

GeoEye 的下一代卫星 GeoEye 2，现在更名为 WorldView 4，最初计划 2013 年年初发射，现在调整为 2016 年年中发射，为第三代精细观测地球卫星，地面分辨率达 0.31 m。

3.4 航天器简介

3.4.1 欧空局(ESA)

为满足地球观测用户群体的需求，从 20 世纪 90 年代中期开始，ESA 着手设计地球观测的双任务，即地球观测任务和地球探测任务。

地球观测任务：主要为预演任务，每项任务包含一系列卫星。主要为了满足特定地球观测应用领域的需求。这种类型卫星的解决方案最终将会转移到可运行卫星的实体上。此类任务的重点在于服务，同时也将解决长期的需求。

地球探测任务：主要为研究或演示任务，每项任务都集中于深入理解地球或大气系统过程。对于某些特定新观测技术的展示也属于这一类型。它们的持续运行时间随着任务类型的不同而不同。

3.4.2 NASA 的航天器

NASA 地球科学事业的任务就是实现科学地理解地球系统以及地球系统对由自然或人类引起的变化的响应，进而提高对气候、天气和自然灾害的预报能力。

地球科学事业已经确定了一系列学术问题的研究策略。其最终的目标就是解答下面的问题：

(1) 全球系统是如何变化的？
(2) 地球系统的主要驱动力是什么？

(3)地球系统对自然和人类引起的变化的响应如何？
(4)人类文明对地球系统变化造成的结果怎样？
(5)我们对地球系统变化的预期能预测到什么程度？

上述五个问题分别定义了地球系统的变化、驱动力、响应、结果和预测等方面的研究途径，并为进一步研究更多更具体的问题指明了方向。

ESE 的空间航天器任务可以归为探测、预操作和技术展示、系观测三类。

1. 探测航天器

探索任务主要用于产生新的学术思想以促进进一步的研究和发展。每颗探索卫星项目设计为一次性的任务，以用于对一些特定的科学问题得出科学结论。

2. 预操作和技术展示

业务化环境代理需要更深入地理解和观测地球系统，同时需要对现有的业务化观测系统进行全面的升级。为了解决上述需求，NASA 投资了一项革新性的传感器技术，发展了性价比更高的、使用更高效的前期科研设备。

3. 系统观测

关键环境变量的系统观测对于研究由于地球系统之外的因素，比如入射太阳辐射的改变等，导致的环境改变至关重要，同时也便于阐述地球系统各主要组成部分的行为特征。系统观测并不意味着一定是连续观测，当在两次间断记录之间短期的自然变量或者校准的不确定性对长期趋势不会造成严重影响时，时间序列存在一定的间隔是可以容忍的。ESE 的目标是系统观测的连续性，因为考虑到传感器技术不成熟或航天器发射可能失败等问题，所以 ESE 并不计划立刻实现连续观测。

下一个十年，NASA 将大量的环境参数从基于研究的计划转移到业务运行状态。这一转移需要对定标、参数反演算法以及数据集的预处理等做细致的规划以保证其一致性，进而保证业务运行卫星的数据获取能力，以至于解决长期的全球变化问题。

4. 长期的稳定观测

AVHRR->MODIS->VIIRS（NPP/NPOESS）

NASA 的地球关系系统(EOS)计划设计为长期稳定观测，主要用于气候研究。3 颗同类卫星持续发射 3 次，可连续运行 18 年以上。

高级的业务化环境卫星计划（NPOESS）提供了类似的观测，组成序列的 3 颗卫星持续跟踪所探测的变量。

从气候研究需求转移到可实际运行系统是可能的，同时也是期望中的、高效的系统，当然也要付出很大努力才能实现。

NPOESS 预备项目（NPP）的初级阶段是利用 MODIS 和 AIRS 观测 14 个参数。

5. EOS 的目标

(1)开展对整个地球系统的理解，以及自然和人类引起的全球环境变化的效应。

(2)利用 NASA 特有的航空、航天以及地面平台能力来探究地球系统有关的学术知识。

(3)发布有关地球系统的信息。

(4)为国家和国际环境政策的制定提供支持。

6. 航天器的主要目标

(1) 创造一个整合的科学观测系统以用于地球系统科学的多学科交叉研究。

(2) 开发一个深入的数据和信息系统，包括数据恢复和处理系统。

(3) 获取和组装全球数据库，并突出持续十年或更长时间的太空遥感观测。

(4) 改善地球系统的预测模型。

7. Terra 的目标

(1) 第一次完成并持续性地进行地球系统诸多特征表面和大气特性的全球拍摄。

(2) 通过人类活动与自然变换相区分的指示物或者痕迹来改进人类对气候影响的探测能力。

(3) 能够探测云、气溶胶和温室气体等影响地球能量平衡的能力。

(4) 提供全球陆地和海洋生产力的评估，以用于全球碳存储、大气交换以及年际变化的精确计算。

(5) 提供的观测数据可以用于改进季节或年度尺度的气候和天气预报的准确程度。

(6) 为改进火灾、火山、洪涝以及干旱等灾害的预测方法、特征以及减少灾害损失。

☞ 小结

本章主要介绍了遥感平台及常用的不同分辨率的遥感数据。

☞ 主要知识点

- 遥感平台
- 轨道和扫描带
- 低、中、高分辨率遥感数据

☞ 课后作业

(1) 请查阅网站 http://landsat.usgs.gov，调查 Landsat 8 和 Landsat 7 光谱波段的对比，并于下次课做个人报告。

(2) 请查阅中国的主要卫星遥感平台和遥感数据，并于下次课做个人调研报告。

第4章 影像数据获取

遥感传感器可以根据其涉及波段的数量及其所探测的频率范围进行归类。通常传感器分为全色、多光谱、高光谱和超光谱传感器。

全色传感器：覆盖了可见光或者近红外波段整个光谱范围。其中黑白相机就是一个只具有一个波段的传感器。

多光谱传感器：具有两个或多个波段的传感器，光谱范围为 0.3~14 m。

高光谱传感器：其光谱波段要比多光谱波段窄很多。具有成百个通道且同时探测某一地物，具有较高的光谱分辨率（图 4.1）。

超光谱传感器：处于研究中，未投入使用。这些传感器具有成千个波段，其波段宽要远远小于高光谱的波段宽。

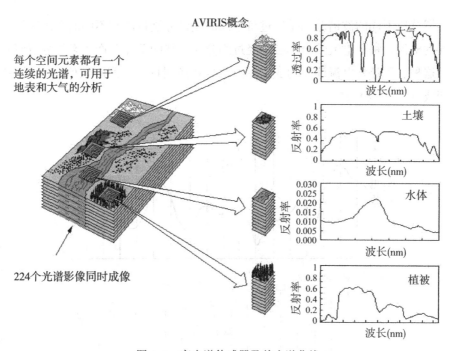

图 4.1 高光谱传感器及其光谱曲线

遥感成像的传感设备可以分为两类：光机扫描仪和电荷耦合设备。

4.1 光机扫描仪

光机扫描仪是通过物理的移动镜面或焦距来系统地观测地表。当设备扫描地表时,通过变化的电流来表征和记录地表地物亮度的差异(图4.2)。

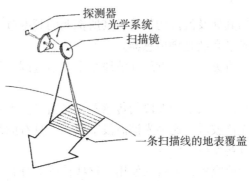

图4.2 光机扫描仪

传感器感应不同的波谱范围是通过将不同的能量分别过滤到不同的光谱区域,每个区域是不同的电流。每个电流信号都必须经过细分为独特的单元并产生影像分析所必需的离散数值。这种过程是将电流连续的变化模拟信号转换到特定的间隔数值(模拟信号向数字信号转变或A-D转换)(图4.3)。

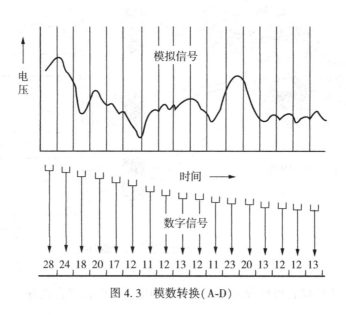

图4.3 模数转换(A-D)

4.2 电荷耦合设备（CCD）

一个 CCD 设备是将感光的物质嵌入到硅片中。势位井接收景物的光子，一般通过一个光学系统来收集、过滤和聚焦辐射予以实现。CCDs 的感应组件可以做到很小，其直径可能小于 $1\mu m$，对可见光和近红外辐射敏感。

这些元素可以通过微电路相互连接来形成矩阵。设计成单独一行的探测设备形成一个线阵，设计成几行几列的探测设备形成二维面阵。

CCDs 可以放置在聚焦平面上，这样它们就能够以与飞行路线正交来探测一个较窄的矩形条带。飞机或卫星的前向运动使得其视场角沿着飞行路径前行，获取其覆盖面的数据。

利用类比方法将这种成像方式称为推扫，而机械扫描则被称为摆扫，其影像的形成依赖于传感器从一侧到另一侧进行移动扫描来实现。

信噪比。每个传感器都会记录与目标物的亮度无关的信息，即噪声，部分是由传感器各个部件累积的硬件误差造成的。为了传感设备的高效利用，设计时其噪声必须要低于其信号。通常使用信噪比来描述（S/N 或 SNR）（图 4.4）。

图 4.4 的底部为一个假想的场景，由两类地物组成。其信号显示，这两类地物之间亮度差别很小。大气效应，传感器误差以及其他因素都是噪声，与信号混合在一起。因此，传感器所记录的是信号和噪声的混合。当相对于信号，噪声很小时（左侧：高信噪比），传感器所记录的信号与噪声差别明显。当相对于噪声，信号很小时（右侧：低信噪比），则传感器就很难区分这两类不同的地物。

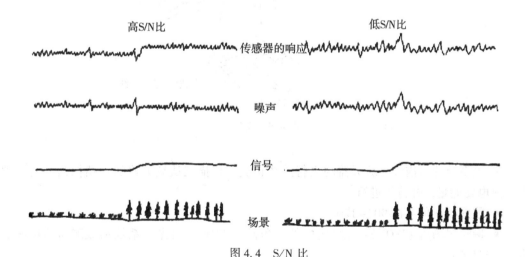

图 4.4 S/N 比

4.3 扫描系统

光电和光谱成像扫描仪通过探测设备探测电磁能量反映出的亮度生成数字影像。根据扫描仪类型的不同，扫描仪可以安装一个或多个感应探测器。

4.3.1 摆扫式扫描仪

摆扫扫描仪为扫描仪的一种类型，也称交叉扫描仪。这种扫描仪使用旋转的镜片在垂直于传感器平台前进的方向从一侧到另一侧扫描地面景物，就像扫帚一样(图4.5)。扫描的宽度称为传感器的扫描带宽度。旋转的镜片将反射的光线聚焦到传感器中心。带有旋转镜片的摆扫式扫描仪形体较大且建造比较复杂。镜片的旋转会使得其扫描的影像发生空间扭曲，因此需要用户在使用前进行预处理。

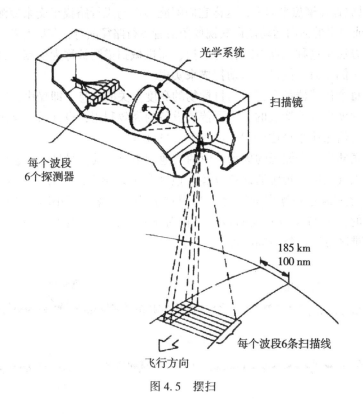

图4.5 摆扫

- 该系统的扫描线与航天器的飞行路线正交。扫描仪从航天器的一端扫描到另一端来探测电磁能量，并持续进行
- 数据收集角度在 90°~120°
- 瞬时视场角 (IFOV)：能量集中的三角锥，IFOV 是由光学系统和探测器的尺度决定的(图 4.6)
- 在扫描市场内，既可能有混合像元，也可能存在纯像元

由图4.6可知，越靠近影像的边缘，像元所占地面的范围越大。小的 IFOV 意味着探测的地面信息越详细，而大 IFOV 意味着：

- 探测器探测的总体能量越多
- 因为较高的信号水平，探测场景的辐射越多

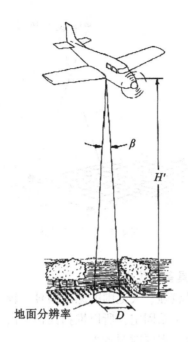

图 4.6 IFOV 和空间分辨率

D 为地面可视圆形区域的直径（空间分辨率）；β 为瞬时视场角；H' 为飞行高度。且 $D = H'\beta$

- 辐射分辨率提高
- 信号比背景噪声要强
- 更高的信噪比
- 更长的滞留时间

4.3.2 推扫式扫描仪

又称沿路径扫描，是另一种类型的扫描仪，不需要使用旋转镜片。传感器探测设备排列成一行称为线阵，进行以一种类似扫帚扫描的形式探测信息。与摆扫从一端扫描到另一端不同，此类传感器利用一维的扫描线阵一次性获取整条线的信息，就像扫帚扫地一样（图 4.7）。近期发射的一些扫描仪均设计为步进凝视型扫描仪，每个波段设置为多行多列的二位阵列。

推扫式扫描仪体型很小，重量很轻，因为其不需要设计移动的机械，所以相对于摆扫传感器来说，推扫式扫描仪的设计要简单一些。同时，此类传感器还具有比较好的辐射和空间分辨率。

推扫式扫描仪一个较大的缺点是需要对组成传感器系统的大量的探测器进行逐一标定。

探测器的规模决定了每个像元分辨率的大小。

每个光谱波段或通道需要生成其独特的阵列。

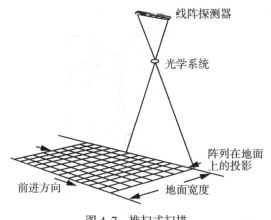

图 4.7 推扫式扫描

1. 相对于摆扫，推扫所具备的优点
(1)滞留时间较长、信号较强、大范围的感应信号、较好的空间和辐射分辨率。
(2)因为探测器元件之间关系固定，所以推扫式扫描仪具有较好的几何关系。
(3)设备较小、重量较轻，需要能源较少。

2. 相对于摆扫，推扫的缺点
(1)需要标定多个探测原件。
(2)商用 CCD 光谱敏感性的范围有限。

4.4 遥感数据

4.4.1 栅格或格网数据模型

像元是影像上可识别的最小单元。形状规则，多为方形的、矩形的或多边形的。其中，方形用的最多。利用影像矩阵(行列形式)来组织地面地表特征。可以探测反射光的大小等物理现象。数值的大小代表的是每个像元信号的强弱。

对于遥感来说，探测的是像元所覆盖的地面范围内的平均电磁辐射。据其平均电磁辐射强度的大小，每个像元赋予一个数值。低或无亮度以最低值来表示，高亮度以最大值来表示，其他的拉伸至最大与最小值之间。

4.4.2 数字影像数据

- 数字数据是由数字(DN 值)组成的矩阵(图 4.8)
- 每个波段为一层(或者说一个矩阵)
- 一个像元具有一个 DN 值
- 影像是 2 维矩阵。每个像元有其位置和 DN 值
- 像元的位置通常用行和列来描述

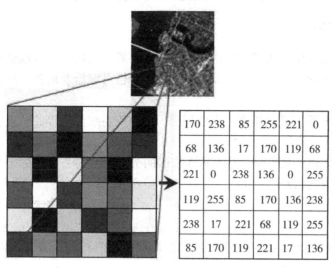

图 4.8　栅格模型和 DN 值

1. 何谓 DN 值

DN 值代表地表辐射量信息的相对大小，而且 DN 值不是反射率。在已知大气状况的前提下，DN 值可以转换为反射率，DN 值的范围取决于传感器的辐射分辨率。

2. 影像的文件格式

影像的格式很多，一般的，包含一个头文件和数据文件。

影像数据的组织可以有不同的形式。如波段序列（BSQ）、逐波段逐行（BIL）和逐波段逐像元（BIP）。

4.4.3　数据获取

- 探测器所探测到的地物反射或发射的电磁波谱可以用数字的或模拟的数值来记录
- 照片的胶卷就起到了探测和记录介质的作用
- 数字传感器产生的电信号与地面景观辐射的能量相对应
- 信号可以转换为影像或者照片

4.4.4　数据解译

- 遥感数据解译主要是为了影像数据的分析
- 照片数据的可视化解译是长期以来遥感的最基本的工作
- 可视化技术充分利用人类出色的影像解译能力来定量地评估影像的空间模式
- 解译元素包括形状、尺寸、图型、色调、纹理、阴影、位置、布局以及分辨率等
- 可视化解译技术存在一定的不足，如需要广泛的训练、劳动强度大等
- 可视化解译中没有充分利用光谱特性
- 因为人眼识别色调的能力有限，很难同时分析多个光谱影像

- 主要应用于光谱模式相对固定的地物的识别，多用于数字数据而非照片

4.4.5 参考数据

遥感数据通常需要一些参考数据。参考数据一般是与遥感所获取的地物、区域或现象数据相对应的地面的真实数据，比如，某一特定分析可能需要用到土壤图、水质测试报告或者航空照片等。

参考数据多来源于对农作物、土地利用、树种或者水污染等的位置、面积以及状况等问题的野外调查。

参考数据的作用主要包括：(1)遥感数据的分析和解译；(2)标定传感器；(3)验证从遥感数据中提取的信息。

☞ 小结

本章主要介绍了光机扫描仪和CCD，摆扫和推扫两种扫描方式，同时对数字数据的格式及其存储进行了阐述。

☞ 主要知识点

- 光机扫描仪和CCD
- 信噪比
- 摆扫和推扫的优缺点
- 栅格数据模型和存储格式
- DN值的含义

☞ 作业

对遥感图像处理软件进行调查，并请同学做汇报。

第 5 章　影像校正和预处理

5.1　预处理

影像的预处理是影像分析的前期工作。一般来说,预处理包括:
几何预处理:将影像与地图或者其他影像配准。
辐射预处理:调整数字数值以消除或减弱恶劣天气对辐射亮度的影响。
经过校正完成后的影像数据即可用于遥感分析。
预处理是一个准备阶段,从原理上来说是提高图像的质量以用于后续的分析。
预处理包含很多工作。具体来说主要包括如下三个步骤:(1)特征提取;(2)辐射校正;(3)几何校正。

5.2　几何校正

1. 何为几何校正

引起像元空间特性变化的任何过程都叫几何校正,比如:像元坐标的变化、像元之间的关系和像元大小的变化等。

几何校正同时也改变像元的辐射量的大小,这主要是由于几何校正时的重采样引起的。

2. 几何校正的意义

(1)实现影像与地图的叠加。
(2)消除由于地形、传感器的不稳定,地球曲率等引起的影像形变。
(3)改变影像的空间分辨率。
(4)改变影像的地图投影。

3. 影像与地图拟合的两种策略

(1)利用地面控制点(GCP)将真实坐标赋给影像来实现影像的校正。
(2)在两景影像或者影像和地图之间进行配准(影像配准)。
需要指出的是这两种策略的原理是相同的。

5.2.1　利用地面控制点进行校正

1. 目标

将影像像元位置与它们在地球上实际的地理位置进行配准(图 5.1)

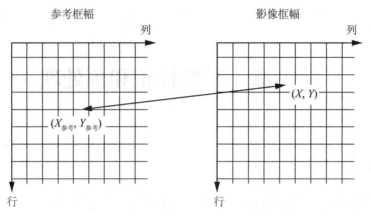

图 5.1 影像校正

2. 方法

(1) 对影像上已知位置赋三维坐标即地面控制点。

(2) 创建一个模型来拟合所有的地面控制点。

(3) 对影像进行变换使得最大程度上与上述模型相匹配。

3. 地面控制点（GCP）的选取

控制点是地图上具有较高精度的特征地物，同时其位置在影像上也比较容易找到。

如果特征点能够从背景中很容易区分的话，那么最理想的效果是控制点要达到一个像元大小。

在实际工作中，大部分控制点一般有几个像元大小。可能是高速公路的交叉点、可区分的水体、土地覆盖地块的边缘、小溪的结合处及类似特征。

通常，找到一个小尺度或中等尺度的控制点比较容易。然而，在很多情况下，分析人员发现很难扩展控制点集，因为每新增一个控制点，其精度就越来越难保证。

因此，好的控制点集包含的点数可能比较少。好的控制点意味着不仅在影像上而且在地图上都要确保其有很好的精度。

控制点的布局也同样是一个大问题。理论上，控制点应该在整个影像中均匀分布，在影像边缘处要覆盖得比较好。

选择好的点同时又要均匀分布，二者很难同时被顾及。对于分析人员来说二者之间很难平衡。

根据 Bernstein 等(1983)的研究，随着控制点数的增多，配准误差降低。因此可以说，控制点数越多越好。

如果每个控制点的精度都在 1/3 个像元以内，那么选取 16 个控制点比较理想。当然，如果这些点的布局相对集中或者地表的自然特征很难保证精度的话，16 个控制点还是不够的，需要再增加点数。

总之，控制点的选择应做到：

(1)在道路交叉点、河湾、自然的特征地物等处选择控制点。
(2)控制点应该布满整个影像。
(3)控制点的最小数量取决于后期校正所使用的转换模型。
(4)最好具备一定数量的控制点集。
(5)同时,必须为控制点坐标系选择一个地图投影。

5.2.2 影像配准与影像校正的差异

- 与影像校正相比,影像配准不是从地图上选择控制点,而是将两个或多个影像关联起来
- 可以用于配准一个没有参考系统的影像
- 可用于将两景影像进行对照(比如在不同时期获取的同一地区的影像等)

5.2.3 影像与地图的拟合过程

变换:利用数学方程将像元坐标与所有的控制点实现很好的配准。

重采样:一旦将像元移动到一个新的位置就需要通过重采样来重新为其赋 DN 值。

1. 数学变换

数学变换即在具备控制点的前提下,在原始坐标与校正后坐标之间建立回归关系,然后将这一关系应用到整个影像数据。转换时可以采用 1 次、2 次或 3 次多项式变换。1 次变换为线性变换,一般用于面积较小的平坦地区,至少需要 3 个控制点。2 次、3 次或更高次变换为非线性变换。2 次变换适合于需要将地球曲率考虑在内的较大面积区域,或者在地形上具有中等变化程度的地区,至少需要 6 个控制点。而 3 次变换则最少需要 10 个控制点,一般适宜于地形变化幅度较大的区域。

通常,控制点的最小数量是 3 的倍数。

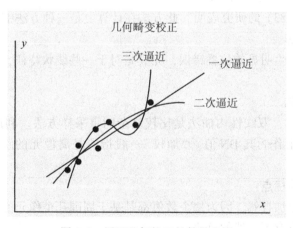

图 5.2 用于几何校正的数学变换

2. 影像重采样

一旦一景影像发生了畸变,那么如何为偏移了的像元赋予其 DN 值呢?这就需要影像

的重采样技术，主要包括：

(1)最近邻：用最近的像元值赋给新像元。

(2)双线性：用最近4个像元值的平均值赋给新像元。

(3)三次卷积：用最近的16个像元数值的平均值作为新像元的DN值。

3. 最近邻法

利用最近邻的未校正的像元的数值赋给校正了的像元(图5.3)

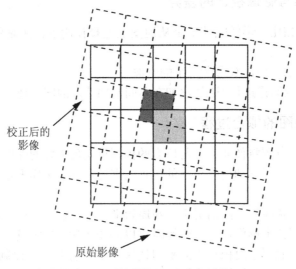

图5.3 利用最近邻法进行插值

最近邻法的优点

(1)简单且能够保留影像的原始数值。

(2)Kovalick(1983)的研究表明，此方法在计算上是三种方法中最高效的方法。

最近邻法的缺点

最近邻法可能产生明显的位置错误，尤其是对于一些线状特征，重采样后像元之间错位明显。

4. 双线性内插法

与最近邻法相比，双线性内插法是比较复杂的重采样方法。利用周围四个最近邻像元数值的加权平均值作为其DN值。"加权"一般指的是离像元的距离越近，权重越大。(图5.4)

双线性内插法的优点

校正后的影像更加自然，因为每个数值都是基于周围几个像元值加权计算得到的，与最近邻相比，此法没有不自然的色块。

双线性内插法的缺点

(1)原始影像中的亮度值丢失了。

(2)因为重采样是利用周围几个像元值平均得到的，因此降低了影像的空间分辨率，存在模糊的块状像元。

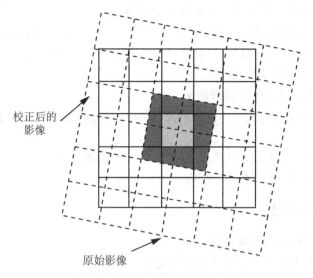

图 5.4 双线性内插法进行重采样

5. 三次卷积内插法

三次卷积内插法最繁杂、最复杂,也可能是使用最广泛的插值方法。这种方法采用周围 16 个像元的加权平均值作为该像元的数值(图 5.5)。

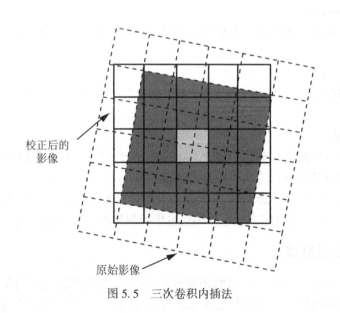

图 5.5 三次卷积内插法

三次卷积内插法的优点
此方法的结果比其他方法的结果具有更好的精度,因此更有吸引力。
三次卷积内插法的缺点
(1)相对于最近邻和双线性,校正后的数值改变较大。

(2)计算强度最大，需要的最少控制点数增多。

为了保持影像的辐射测量数值，做光谱分析一定要采用最近邻重采样方法。如果需要得到平滑的影像，最好采用双线性或者三次卷积内插法。需要永远记住的是，无论出于何种原因进行重采样，影像的数据都会发生改变。

6. 重采样改变了影像的空间分辨率

(1)提高分辨率（人为地缩小像元的尺寸）：仅仅是将原始像元的 DN 值赋给落在其范围内的更小像元。

(2)降低分辨率(人为地增大像元的尺寸)：以某种方法(比如平均)来合并原始像元的 DN 值，并将其赋给新的较大的像元。

5.2.4 几何校正小结

- 几乎对于所有的遥感项目都是必要的
- 将影像与现实世界的坐标系对应起来
- 对影像和 GIS 的结合起到关键作用
- 对于获取精确的空间产品很必要

5.3 辐射校正

1. 何为辐射校正

辐射校正是指任何改变像元 DN 值的过程，比如大气校正、对比度拉伸、滤波、条带去除等。

2. 何为大气校正

大气校正指的是将在卫星上观测到的辐射量转换成地表的反射率或者出射辐射的方法。

3. 大气影响

太阳辐射的电磁波谱在到达卫星之前需两次经过大气，组成大气的分子与电磁辐射发生相互作用，多数情况下，大气效应是与波段相关的。

大气对电磁辐射的主要影响有 2~3 种，主要包括大气散射、大气吸收(或者相反的作用——透过等)，大气也可能造成电磁辐射发生折射即改变其前进方向。

5.3.1 大气校正概述

任何利用可见光或近红外的辐射来获取地表信息的传感器记录的都是两类辐射亮度值。一类是地表的反射率，这部分是遥感最为关注的；另一类就是大气本身的辐射亮度即大气散射的影响。

因此，传感器所观测到的亮度是反射率和大气散射作用的综合结果。我们不能立即将这两类亮度区分开来，而大气校正的目标之一就是鉴别和分离这两类辐射亮度以便于遥感分析的对象能够主要集中于监测地表的辐射亮度。

很多大气校正方法都是尽力去除路径辐射对总辐射的影响。这要假定对于整幅影像，其大气状况类似。同时需要逐个波段分别进行大气校正，因为绿波段的大气辐射是近红外波段大气辐射的4倍，而与近红外相比，可见光部分的大气效应更加强烈。

5.3.2 路径辐射校正

路径辐射校正必须逐波段进行，并且必须估计路径辐射的数值并将它去除掉。

在技术上主要包括"暗像元去除法"或者利用短波与未发生散射作用的长波做回归分析来实现。

1. 暗像元法

假设影像中存在暗像元，其 DN 值为 0，即不发生发射作用，没有反射率（当然这种假设并非总是成立）。

利用直方图等方法找到波段中像元的最小 DN 值（图 5.6）。

该波段中所有其他像元的 DN 值减掉这个最小值即实现了路径辐射的校正。

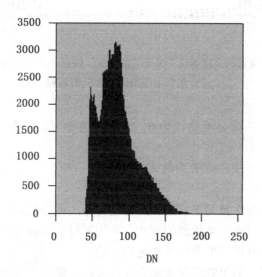

图 5.6 利用直方图查找最小 DN 值

2. 暗像元法实例

首先，在场景中找到一个暗像元地物。比如大面积的水体或者云的阴影或者高大陡峭的山体阴影等。

在光谱的红外区域，水体和阴影的亮度值都接近 0，因为干净的水体在近红外波段具有强烈的吸收，对阴影区域而言，只有非常少的能量被阴影散射到传感器上。

分析人员检测上述这些区域或者利用 DN 值直方图时，发现水体等的暗像元的最低值并非是 0，而是一些较大的数值。

通常，暗像元 DN 值的大小在各个波段之间存在着差异，比如 Landsat 卫星 TM 数据，暗像元的 DN 值在第一波段可能是 12，在第二波段可能是 7，在第三波段可能是 2，在第四波段可能是 2。这些数值，可以认为是大气散射作用的结果。因此，每个波段每个像元的 DN 值都减掉这一波段暗像元的 DN 值，即消除了大气辐射对总辐射的影响，实现了大气校正(图 5.7)。

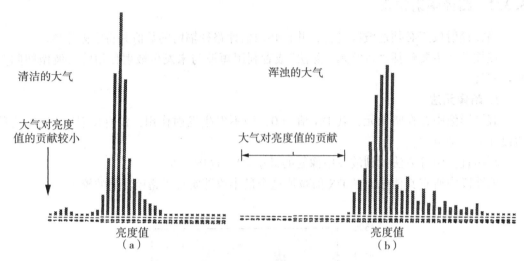

图 5.7 最低的亮度值认为是大气的贡献，因此波段中所有像元减掉
这个最低值即消除了大气的影响

因此，每个波段的最低亮度值设为 0，则深黑色即可认为是黑色地物无大气散射的结果。

这个方法是最简单、最直接的去除大气影响的一种方法(Chavez，1975)，有时称为(最小直方图法)(HMM)。

此过程方法简单易行，且易于应用，因为这种方法充分利用了影像自身的信息。

当然，这一方法仅仅是一个近似的方法。大气不仅影响像元空间的位置而且影响其形状，即大气对影像中所有像元的影响并非是均一的。

此外，在干旱地区，如果观测时太阳高度角很高，那么影像中阴影、云以及大面积水域将非常稀少，即使有，分布面积也很小。因此，在这种情况下，此方法的应用存在很大难度。

5.3.3 大气透过率校正

- 每个波段单独进行
- 必须假定或者测量光学厚度、各大气组成成分的密度等
- 因为其很复杂和困难，所以通常不做大气透过率校正

5.3.4 大气测量和建模

当卫星通过时，需要同步测量不同高度处的大气特性。

目前已有一些封装好的成熟的大气模型，如 LOWTRAN、MODTRAN。

LOWTRAN 能够计算各种大气条件下的大气透过率和大气背景辐射。LOWTRAN 适合

于具有较低光谱分辨率的影像,而 MODTRAN 则适合于光谱分辨率较高的影像。

这些程序利用大气中的气体分子估算大气的吸收率和发射率。

这些模型累计不同的大气条件,包括季节和地理的变化,云的状况、雨以及阴霾天气等。这些模型需要考虑尽可能多的大气路径。

5.3.5 大气校正的意义

大气校正并非总是必需的,比如在对单景影像研究时,或者大气的影响通过植被指数比率如 NDVI 等消除时,可以不用考虑大气校正。

通常在对比多景影像时,需要进行大气校正,比如进行影像配准(镶嵌)、变化检测或者将分类统计结果应用到多景影像时。

如果需要计算地表的反射率或者需要对比星上和地面上观测的反射率时,大气校正是必须要做的一项工作。

5.3.6 辐射校正小结

- 改变传感器获取的像元的 DN 值
- 通常是去除辐射如路径辐射而非直接去除目标
- 因为改变了原始的数值,所以校正时需谨慎

5.4 特征提取

在影像处理中,特征提取(或特征选择)一词具有特殊的意义。"特征"并非指在影像上可见的地理特征,而是影像数据的统计特征:即单个波段或者波段组合的数值反映了影像所对应场景的系统变化。因此,特征提取也称为"信息提取",在描述影像的统计信息方法中,多光谱数据的组分分离技术是用得最多的一种。在理论上,被舍弃的数据包含了原始数据中的噪声和误差。因此,特征提取可以提高精度。此外,特征提取减少了需要分析的光谱通道或者光谱波段的数量,降低了对计算性能的要求。经过特征提取之后,分析人员就可以将主要精力用于少量但携带信息更丰富的波段,这些波段数据所含有的信息几乎与全部波段数据所含的信息一致。此外,特征提取还会提高速度,降低分析成本。

多光谱数据包含多个通道的数据,少则包含 3 个、4 个或者 7 个,多则 200 个或者更多。对于这么多通道数据的处理,即使是中等尺度的影像也需要很长时间。从这个意义上说,特征提取有很好的实际意义,因为影像分析人员希望在保持影像数据的高效、精确的基础上,通道数据越少越好。

下面的例子是基于 TM 数据。TM 数据具有 7 个通道,足以说明特征提取这个概念,同时也足够精确。一个方差和协方差矩阵可以用来说明两波段之间的关系。有些相关性很强,比如波段 1 与波段 3,波段 2 与波段 3,它们的相关系数都在 0.9 以上。相关性越高,说明这两个通道数据的关系越紧密。因此,当通道 2 的数据升高或者降低时,通道 3 的数据也会发生同样的变化。即一个通道与另一个通道的信息相近。特征提取的目的就是首先鉴定信息相近的通道,然后去除其中之一,以便于用最少的波段数来表达尽可能多的信息。

从表 5.1 可以看出，波段 3，5 和 6 所含的信息几乎可以代表整个 7 个波段的信息，因为波段 3 与波段 1 和 2 密切相关，波段 5 与波段 4 和 7 密切相关，而波段 6 所含的信息与其他波段的信息相差很大。因此，被舍弃的波段(1，2，4 和 7)均被分别整合到保留的三个通道中。可以看出，特征提取通过舍弃没必要的波段，达到减少波段数的目的。虽然这种方法可以作为最基础的一种特征提取途径，但是一般来说，特征提取是一种基于在波段间进行更复杂的相关性统计的过程。

表 5.1 **TM 数据 7 个波段的相似矩阵协方差矩阵**

	1	2	3	4	5	6	7
1	48.8	29.2	43.2	49.9	76.5	0.9	44.9
2	29.2	20.3	29.0	48.6	65.4	1.5	32.8
3	43.2	29.0	46.4	59.9	101.2	0.6	53.5
4	49.9	48.6	59.9	327.8	325.6	12.4	104.3
5	76.5	65.4	101.2	325.6	480.5	10.2	188.5
6	0.9	1.5	0.6	12.5	10.2	14.0	1.1
7	45.0	32.8	53.5	104.3	188.5	1.1	90.8

相关性矩阵

	1	2	3	4	5	6	7
1	1.00						
2	0.92	1.00					
3	0.90	0.94	1.00				
4	0.39	0.59	0.48	1.00			
5	0.49	0.66	0.67	0.82	1.00		
6	0.03	0.08	0.02	0.18	0.12	1.00	
7	0.67	0.76	0.82	0.60	0.90	0.02	1.00

主成分分析法(Principal Component Analysis，PCA)是最常用的特征提取的方法之一。本章只对其进行简单的描述，详细的介绍请见文献 Davis(1986)或者其他。本质上来说，PCA 寻找原始波段间的最优线性组合来表达影像中像元值的变化。线性组合的基本形式为：

$$A = C_1X_1 + C_2X_2 + C_3X_3 + C_4X_4$$

其中，X_1，X_2，X_3 和 X_4 是四个光谱波段的像元值，C_1，C_2，C_3 和 C_4 是各光谱波段的系数。A 是像元转换后的数值。举例来说，系数 $C_1 = 0.35$，$C_2 = -0.08$，$C_3 = 0.36$，$C_4 = 0.86$，四个波段某一像元的像元值 $X_1 = 28$，$X_2 = 29$，$X_3 = 21$，$X_4 = 54$，那么这个像元转换后的数值 A 为 61.48。系数的确定需要假定通过这些系数计算得到的数值表征了整个影像的最大变化。这样，通过系数对原始波段的线性组合可以产生一个含有最多信息的单个波段。如果将这一过程运用到整个影像中的所有像元,那么所产生的波段含有的信息几乎与影像原始的四个波段的信息一致。

当然,PCA 计算结果是否有效取决于最优化系数。本章只是缩略的描述,因为系数的计算方法非常复杂,比如高等数理统计方法或者比如 Davis（1986）和 Gould（1967）所描述的方法。我们需要掌握的是,PCA 的关键之处在于其最优化系数的确定,要保证系数能够最大程度地浓缩单个波段的信息。

此过程还会产生第二组系数,并由其产生第二组像元值,这里记为 B 组系数或者 B 影像。虽然系数即影像信息的有效性要低于第一组,但是也能代表影像中像元值的变化。这一过程最终要产生 7 组系数和 7 景影像值,分别记为 A、B、C、D、E、F 和 G,每个都比其前一个所含的信息要少。表 5.2 中,Ⅰ 和 Ⅱ 的信息之和占了影像总变化的 93% 左右,而 Ⅲ～Ⅶ 的信息之和则只占到 7% 左右。为了减少波段数量,分析人员很有可能舍弃这 7% 的信息。即使如此,还有 93% 的信息予以保留,这也足够其用于分析。可见,特征提取通过消除重复的信息而达到缩减数据的目的。

表 5.2　　　　　　　　　　根据表 5.1 数据得到的主成分分析表

	成　分						
	Ⅰ	Ⅱ	Ⅲ	Ⅳ	Ⅴ	Ⅵ	Ⅶ
	特　征　向　量						
变量%	82.5%	10.2%	5.3%	1.3%	0.4%	0.3%	0.1%
特征值	848.44	104.72	54.72	13.55	4.05	2.78	0.77
	0.14	0.35	0.60	0.07	-0.14	-0.66	-0.20
	0.11	0.16	0.32	0.03	-0.07	-0.15	-0.90
	0.37	0.35	0.39	-0.04	-0.22	0.71	-0.36
	0.56	-0.71	0.37	-0.09	-0.18	0.03	-0.64
	0.74	0.21	-0.50	0.06	-0.39	-0.10	0.03
	0.01	-0.05	0.02	0.99	0.12	0.08	-0.04
	0.29	0.42	-0.08	-0.09	0.85	0.02	-0.02
	加载某点数值						
Band 1	0.562	0.519	0.629	0.037	-0.040	-0.160	-0.245
Band 2	0.729	0.369	0.529	0.027	-0.307	-0.576	-0.177
Band 3	0.707	0.528	0.419	-0.022	-0.659	-0.179	-0.046
Band 4	0.903	-0.401	0.150	-0.017	0.020	0.003	-0.003
Band 5	0.980	0.098	-0.166	0.011	-0.035	-0.008	-0.001
Band 6	0.144	-0.150	0.039	0.969	0.063	0.038	-0.010
Band 7	0.873	0.448	-0.062	-0.033	0.180	0.004	-0.002

PCA 的效果如图 5.8 所示。图 5.8 是 TM 影像经过 PCA 处理后的结果。影像 PC I 和 PC II 信息最丰富，而 PC III、PC IV、PC VI 和 PC VII 所含信息逐渐减少，这是因为影像中记录的系统噪声、大气散射以及其他与地表无关的信号造成的。如果在后续的分析中舍弃这两个通道，那么在减少计算的同时，也能保证分析结果的精度。

虽然 PCA 不是特征提取的唯一方法，但是也能够阐明特征提取的目的，即在保证有用信息的前提下，减少用于分析的通道数量，以减少噪声和误差对分析结果造成的影响。

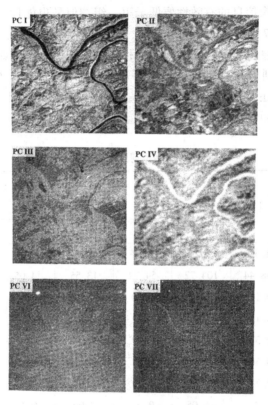

图 5.8 特征提取

影像为表 5.1 和表 5.2 所描述的经过 PCA 处理后 7 个主成分中的 6 个。第 1 主成分（PC I）是由所有 7 个波段数据线性组合形成的，占了影像总变化的 82.5%。PC II 和 PC III 分别占了 10.2% 和 5.3%。更高的主成分即 PC VI 和 PC VII 占了影像总变化的一小部分，主要为噪声和误差，这可以从影像模式中显示出来

5.5 图像裁剪

因为遥感影像一般尺寸很大，而分析人员手头的项目又仅仅是一景影像的部分区域。所以，为了节省计算机的存储空间和分析人员的时间、精力，每个项目开始的第一项任务是图像裁剪，即从整景影像中裁剪出感兴趣的工作区。

尽管对于从事遥感影像解译的人来说，图像裁剪并非难事，但是也并不容易。通常，

在图像裁剪之前，首先需要将图像与其他数据进行配准，这就需要在影像和数据中寻找特征地物以确保二者在空间上的一致性。其次，影像越大，影像与地图或其他影像配准所需要的时间越多，付出的努力越大，因此在影像配准之前最好进行图像裁剪。

在影像裁剪时还需要考虑后续的分析。裁剪子集的区域应该足够大，以确保特定分析之用，比如，在影像分类时，子集要足够大，以便能够选择足够数量的训练区域，或者为分类的精度验证提供足够的样本采集区域。

☞ **知识点**

- 几何校正：地面点的选取，影像匹配，重采样技术
- 辐射校正：大气校正（暗像元法）
- 主成分分析

☞ **思考题**

（1）影像分析人员如何评定影像预处理的结果？

（2）假设某个公司出售已经经过预处理的影像产品，你会对这些产品动心吗？请回答并解释原因。

第6章 影像解译

6.1 简介

有些信息，比如在影像的各种色调和纹理中隐藏的信息，我们是不能直接得到的。

为了将影像转变成信息，我们需要应用特定的知识即影像解译来进行，通过影像解译，我们可以从原始遥感系统所获取的原始影像中获得有用的信息。

6.1.1 影像解译的前提

主题：我们对所解译主题的认知，这也是促使我们进行影像解译的动机，是解译的核心。

地理区域：影响影像记录模式的独特特征。

遥感系统：影响影像解译的变量以及如何估计它们的数值。

6.1.2 影像解译的任务

1. 影像解译的任务——分类

分类：是根据目标、特征或者区域等在影像上的外在显现将其识别出来。

检测：确定某个特征存在与否。

识别：关于特征或物体的更高水平的认知，以便于将物体归类到某个类别中。

鉴别：有确凿的证据和充分的理由将物体或特征归为某个特定的类中(图6.1)。

一般来说，解译人员利用"可能"或者"或许"来鉴别以评估其对所鉴别地物的确定性程度。

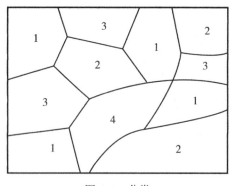

图 6.1 分类

2. 影像解译的任务——枚举

枚举是列举或者清点影像中可见的离散要素(图6.2)。

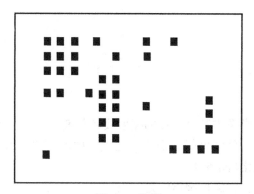

图6.2 枚举

3. 影像解译的任务——测量

首先，距离或者高度的测量，也可能扩展到面积或体积。

其次，对影像亮度的定量估计(图6.3)。

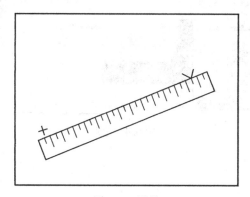

图6.3 测量

4. 影像解译的任务——勾勒

单独区分以不同色调或纹理为特性的不同的面积单元，确定不同区域的边缘或者边界。

6.2 影像解译元素

常用的影像解译的元素主要包括：
- 色调
- 纹理

- 阴影
- 图型
- 布局
- 形状
- 尺寸
- 位置

6.2.1 色调

影像的色调主要指影像中某个区域的亮度或者灰度。

对于黑白影像来说，色调可能标识为"亮"、"中度灰"、"黑灰"、"黑"等，因为影像在白、灰、黑之间变化(图6.4)。

对于彩色影像而言，影像色调指的是"颜色"。

影像解译人员多在黑白影像上利用色调进行解译，彩色影像解译用色调则很少见。尽管人工解译不可能描述影像绝对的亮度值是多少，但是仍然可以区分色调上的差异。

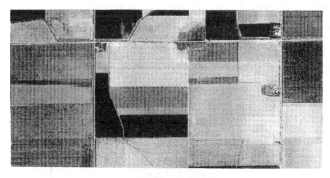

图6.4 色调

6.2.2 纹理

影像的纹理指的是影像色调的空间变化，可以通过平滑的、波纹状、线状的、不规则的或者是斑块状的等来形容影像的纹理。

通常，纹理是由高亮或阴影区域形成的，而高亮或阴影多是由在倾斜的角度上照射不规则表面引起的(图6.5)。

遥感解译人员擅长于区分影像中细微的纹理差异，因此纹理非常有利于影像解译，在很多情况下纹理与色调同等重要。

6.2.3 阴影

阴影为解译物体提供了一个重要的线索。一个建筑物或者车辆，在某个角度照射时，会产生阴影，用这一阴影可以揭示地物的尺寸和形状等。而在光线垂直照射时阴影则不明显。

图 6.5 纹理

阴影在解译人造景观，识别不同结构或物体时很重要。因为军事照片解译人员通常将主要注意力集中于单个设备的识别上，他们开发了很多方法利用阴影来区分细微的差别，而用其他方法则很难识别(图 6.6)。

阴影在自然现象的解译中也具有重要意义。

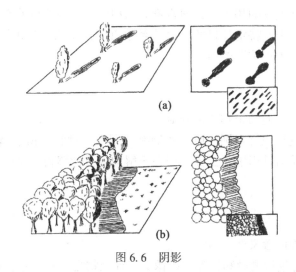

图 6.6 阴影

阴影的重要性：
(1)灌木丛在开放领域中由阴影形成的特征模式。
(2)林区的阴影增强了两个不同土地利用类型边界。

6.2.4 图型

图型指的是由单个物体排列成独特的形式以便于对影像的识别。

影像中的图型中，组成图型的各单体之间存在函数关系(图6.7)。

在土地利用相关研究中，很多模式都与人类的活动密切相关，比如，农田纵横交错的模式，森林采伐线与城市的有序模式以及公路、铁路等的线性模式等。

图6.7 图型

6.2.5 布局

布局指某些特征地物的相关性，通常没有图型反映出的严格的空间排列。

在军用影像解译中，特定项目的依存关系很重要，比如，当识别出某类特定设备时，可能在其附近就有其他更重要的项目。

6.2.6 形状

特征的形状对于它们的识别非常明显。

可以利用形状进行地物识别的典型实例有冲积扇、背斜、植被覆盖区，以及海浪等。之所以可以通过形状对其识别，主要原因在于解译人员能够将所见的形状特征与某类地物紧密关联起来。

6.2.7 尺寸

尺寸有两个方面的含义。

首先，即使没有进行测量和计算，影像上特征地物之间的尺寸差别为解译人员提供了对其尺度和分辨率的识别。

其次，实际的尺寸测量对于解译来说也同等重要。

6.2.8 位置

位置指的是地理位置。比如，污水处理厂需要安置在靠近溪水或河流附近较低的地形区域，以便于收集从较高位置排放的废水。

6.3 影像解译策略

影像解译一般包括如下几个步骤：
- 野外踏勘
- 直接识别
- 推理解译
- 概率性解译
- 确定性解译

6.3.1 野外踏勘

当影像及其与下垫面的关系不清时，解译人员需要进行野外踏勘来勘察地面状况。

分析人员根据现有的知识和经验不能解译影像时，需要收集野外观测的信息来确定影像及其与下垫面的关系。

6.3.2 直接识别

利用解译人员的经验、技巧和辨识力将影像图案和信息类型融合即为直接识别。

直接识别的过程本质上是利用影像判读元素作为可视和逻辑线索，对影像给出一个定性的、主观的分析。

6.3.3 推理解译

推理解译是根据影像上可见的部分推理出在影像上不能直接识别的信息成图的过程。可见信息的分布在此过程中扮演了最终成图的替代品的角色。

比如，遥感影像不能够直接提供关于土壤垂直分布的信息，但是土壤的分布有时与影像上能够反映出的土地形式和植被的图案有密切的关系。

6.3.4 概率性解译

概率性解译是借助于定量分类算法将非影像信息与分类过程进行整合来尽可能缩小概率解译的范围。

6.3.5 确定性解译

确定性解译是确定地面状况和影像特性之间的定量关系。

与其他方法相比，更多的信息都是从遥感影像本身获取的。

这也是影像解译最严格和最精确的一种方式。

☞ 小结

本章主要介绍了影像解译的概念、任务、元素以及解译策略等。

☞ **知识点**
- 影像解译的任务
- 影像解译的主要元素

☞ **思考题**

有些人认为，随着数字解译的出现及其广泛应用，目视解译已毫无用武之地，你是否同意这种看法，请说明理由。

第7章 影像分类

7.1 引言

数字影像分类是遥感科学的基础过程。通过分类，可以将影像获取的像元与地表面的土地覆盖或者土地利用类别联系起来。通常，在分类时，以像元为单元，像元中包含几个光谱波段的数值。通过像元与某个已知类别的像元对比，可以将具有相似像元值的像元归类，并与遥感数据用户所关心的信息类关联起来。这些类别在地图或影像上以区域的形式存在，因此，数字影像经过分类后呈现出多个地块的形式，每个地块以某种颜色或某个符号来表示(图7.1)。理论上来说，这些类别是均一的，即类内像元在光谱上具有很大的相似性，而不同类别之间，像元的相异性大。当然，在实际分类时，类内也呈现一定的变化性，因为对于一景影像而言，即使是相同类别的地物也会呈现出一定程度的差异性。

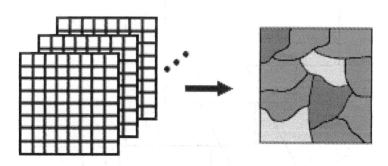

图7.1 影像数据和分类影像

右侧的分类影像是通过检测左侧数字影像，然后将具有相似光谱特性的像元分类后得到的。通常分类影像具有多个类，且需要至少3到4个光谱波段。

影像分类在遥感领域、影像分类和模式识别中具有重要作用。在有些情况下，分析的主要目标就是分类。比如，遥感数据是土地利用分类的主要数据源，并得到了广泛的应用。土地利用分类可以产生类似地图的影像，而这恰恰是分析的最终产品。还有一些情况是，分类后的数据可以作为GIS中的一个图层，以便于进行下一步更精细的分析。比如，在水质的研究中，第一步可能是利用影像分类技术来区分湿地和开放水域。接下来，将对这些区域进行更加细致的研究以便于确定影响水质的主要因素，并制作水质变化图。因此，影像分类是数字影像检测的主要工具，有时是为了制作分类产品，有时是经过分类后从影像中进一步提取信息。

分类器指的是运用特定的过程进行影像分类的计算机程序。目前，科学家设计了很多分类策略，已经发展了很多可用的分类方法，比如，监督的、非监督的、决策树或者是基于知识的、面向对象的、人工神经网络、支持向量机模型以及随机森林等。然而，没有一种方法是十全十美的，每种方法都各有千秋。为了完成特定的任务，分析人员需要在分类之前选择一种合适的分类方法。因此，分析人员需要对影像分类的策略有一个实质的了解才能选择最适合完成手头任务的分类器。

最简单的影像分类是将每个像元作为独立的单元来考虑，根据其在几个光谱波段的数据将其指定到某个类别中(图7.2)。这样的分类器有时被称为光谱分类器或者点分类器，因为这种分类方式将每个像元作为一个点来考虑(即该点的数值与其邻近像元的数值没有关系)。虽然点分类器既简单又经济，但是这种分类方式不能用于揭示在相邻像元中具有相互关联的影像中进行信息的提取。比如，人工解译很难通过一个像元的亮度来获取类别信息，而是通过多组像元亮度的相互关系、呈现出的图型、大小、形状以及相邻像元地块的排列等方法来判定类别。

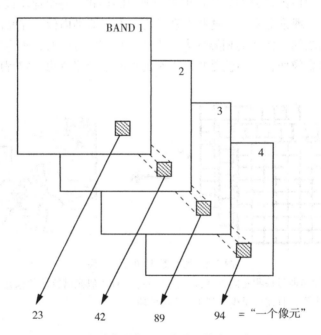

图7.2 点分类器作用于光谱影像中的单个像元

与点分类器相反，大多更为复杂的分类过程需要考虑影像中像元的空间关系，因为对于人工解译来说，像元的纹理信息尤为重要。这些分类器为空间或者邻近像元分类器，它们以影像中一小部分区域为单元考虑其光谱和纹理信息来进行影像分类(图7.3)。这样的空间分类器较点分类器更难设计且费用较高。在有些情况下，空间分类器具有很好的精度，但是对于遥感影像分类来说，能作为常规工具使用的却相对较少。

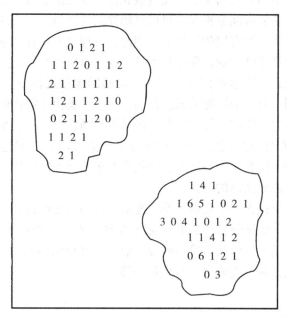

图 7.3 空间分类器基础的影像纹理

空间分类器需要考虑影像某一区域内像元的 DN 值。图中的两个区域是与影像平均亮度值和整体纹理具有明显差异的区域，且在区域内的像元具有较好的一致性

影像分类的另一种方式是将分类方式分为监督分类和非监督分类。监督分类是从遥感数据中提取定量信息的主要工具。要利用监督分类，分析人员需要利用足够多的一种类别的像元点生成具有代表性的参数信息，这个过程即为样本训练。经过训练后，监督分类器就可以利用这些参数对未知像元进行分类。而非监督分类则几乎无需分析人员的干预，主要采用一些聚类算法实现影像的分类。区分监督分类和非监督分类很关键，尤其是对于刚接触影像分类的学生来说尤为重要。但是，这两种分类策略并非泾渭分明，因为有些分类方法并非百分百属于某种分类策略。那些所谓的混合分类器既有监督分类的特性也有非监督分类的特性。

7.2 信息类和光谱类

信息类指的是数据使用用户所感兴趣的类别，比如，不同种类的地层，不同类型的森林，或者不同用途的土地利用等都是信息类，这些信息是规划设计人员、管理人员、行政人员以及科学家所需要的。这些信息是我们期望能够从影像数据中提取出来的信息，也是我们分析的主要目的。不幸的是，这些类别并非直接记录在遥感影像中，我们需要从影像所记录的亮度信息中根据一定的证据间接地分析出来。比如影像中不能直接反映地层信息，影像中更多的是反映地貌、植被、土壤颜色、阴影等的差异，分析人员根据这些差异来断定在某些特定区域存在某种地层。

光谱类指的是在不同的光谱通道中亮度值相近的像元的集合。分析人员可以从遥感数据中检测出光谱类，如果可能分析人员可以将影像的光谱类与用户所感兴趣的信息类对应起来，这样影像就成为一个有用的信息源。因此，影像分类过程就是将光谱类与信息类相匹配。如果匹配的结果正确无疑，那么分类信息就可以供用户使用。如果光谱类和某些信息类很难匹配，那么，对于这些信息类来说，遥感影像很难提供。在信息类和光谱类之间，我们几乎找不到一对一的匹配关系。任何一种信息类由于类内的自然变化使得在影像上的光谱信息发生变化。比如，即使在某一区域的林地，其年龄、生物组成、密度和生物量等可能不同，在"森林"这一信息类中该区域依然是"森林"，但是在影像上却有不同的光谱反映。除此之外，还有其他的因素，即使是对于光谱类非常均一的信息类别，由于亮度和阴影的差异也会造成光谱的变化。

因此，一般信息类包含多个光谱子类，光谱不同的像元组合在一起也可以形成一个信息类(图 7.4)。在数字影像分类中，我们必须要将各光谱子类作为一个独立的单元来分类，但是在最终的分类图中以某一单一符号来表示，并提供给用户使用(毕竟，这些用户仅仅对信息类感兴趣，对分类的中间结果并不感兴趣)。

7.3 非监督分类

非监督分类过程的主要特点是根据影像的特征来对地表覆盖类型进行初步分类，然后再由影像解译人员对分类后的数据确定其具体的覆盖类型。这种聚类方法试图通过自动寻找隐藏在数据下的结构信息，将具有相似光谱特性的遥感数据聚为一类。分析人员仅需要指定聚类的个数即可。

7.3.1 非监督分类的优点

相对于监督分类而言，非监督分类具有如下优点：

(1)非监督分类不需要额外的资源。非监督分类和监督分类所需要的知识不同。要进行监督分类，需要对区域有详细的了解以便于能够在图上选择每一类别的样本。而要进行非监督分类，则无需该区域的先验知识，分类是自动在未知的地表覆盖类型上采样得以实现的。

(2)人为的误差可能性最低。或许需要指定期望分类的类别数(或者类别的最大和最小数目)，有时需要确定每个组的独特性和一致性。因为非监督分类几乎不需要细节信息，因此发生人为误差的可能性很小。如果分析人员对该区域知之甚少，也可以进行非监督分类。

非监督分类为大数据量的遥感数据的快速信息提取提供了一个非常有用的方法。

7.3.2 缺点和不足

非监督分类的缺点主要在于：其一，过分倚重于自然类别；其二，将这些类别与信息类相匹配时存在一定的困难。

(1)非监督分类基于光谱数据所确定的具有光谱一致性的类别并非是分析人员所感兴

趣的信息类。因此，分析人员需要处理非监督分类所确定的光谱类和最终用户所需要的信息类匹配的问题。目前，在两组类别中，简单的"一对一"的关系几乎是不存在的。

（2）分析人员对输出类别的数量及类别的特定属性的控制有限。因此有必要提供一个关于信息类的说明（比如与其他日期或临近区域的类别相匹配的信息），但这样对非监督分类的使用很难让人满意。

（3）由于对地表覆盖类型知之甚少，所以非监督分类的精度可能很难保证。

当先验知识非常有限的时候，非监督分类的作用就会凸显。在多光谱遥感数据分类中，利用非监督分类的主要目的之一就是获取更多的有用信息，以便为监督分类训练区的选取提供支持。

7.3.3 距离度量

图 7.4 显示了绘制在多维数据空间中的两个像元，每个都具有几个不同光谱通道的观测值。为便于说明，这里仅仅给出两个波段，当然多个波段的基本原理也是相同的。

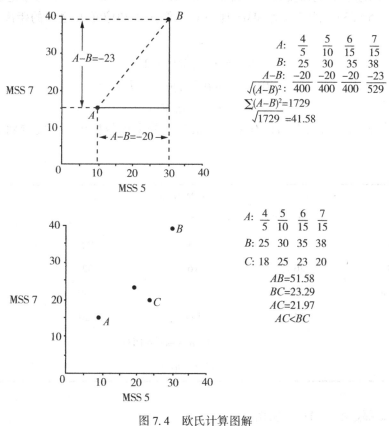

图 7.4 欧氏计算图解

对一景影像的非监督分类需要考虑几千个像元。但是其分类过程仅仅是基于同一个问题的回答，这个问题就是："两个像元属于同一类别吗？"在本例中的问题是"像元 C 是和像元 A 一个类别呢还是和像元 B 一个类别呢?"对该问题的回答可以通过测定两组像元的举例来确定。如果 A 和 C 之间的距离大于 B 和 C, 那么 C 和 B 就属于同一个类别, 而 A 是另一个类别中的一员。

在一景遥感影像中具有成千个像元。如果它们都从属于一定的类别, 那么通常会将某个像元与其他像元的距离作为其分类的依据。那么, 这个距离如何计算呢? 在多维数据空间中计算距离的方法很多, 最简单的一个就是欧氏距离, 公式如下:

$$D_{AB} = \left[\sum_{i=1}^{n} (A_i - B_i)^2 \right]^{1/2} \tag{7.1}$$

其中, i 表示 1 到 n 个波段, A 和 B 为像元, D_{AB} 为像元 AB 的距离。这个距离的计算可以根据勾股定理得到:

$$c = \sqrt{a^2 + b^2} \tag{7.2}$$

在本例中, 我们感兴趣的是距离 c。a, b 和 c 的单位依据两个光谱通道的单位而定。

$$c = D_{AB} \tag{7.3}$$

为了计算 D_{AB}, 我们需要找到距离 a 和 b。距离 a 为 MMS 通道 7 的像元 B 减去像元 A 得到 ($a = 38-15 = 23$)。距离 b 为 MMS 通道 6 中像元 A 和像元 B 之间的距离 ($b = 30-10 = 20$)。那么:

$$D_{AB} = c = \sqrt{20^2 + 23^2}$$
$$D_{AB} = \sqrt{400 + 529} = \sqrt{929}$$
$$D_{AB} = 30.47$$

这种计算方法可以应用到具有多个光谱通道的多维数据中, 只需要将各距离累加。比如:

	Landsat MSS band			
	1	2	3	4
Pixel A	34	28	22	6
Pixel B	26	16	52	29
Difference	8	12	-30	-23
(Difference)^2	64	144	900	529
Total of (differences)^2 = 1637				
$\sqrt{\text{total}}$ = 40.5				

图 7.4 底部显示了另一个实例。

这样, 像元 A 和 B 之间的欧氏距离为 40.45。这个数值本身没有意义, 但是可以将其与其他距离数值进行对比, 并以此作为衡量像元之间相似性的指标。假设, 我们发现 AB 之间的距离为 40.45, 而 AC 之间的距离为 86.34, 那么据此我们断定, 与 B 和 C 相比较, 像元 A 与像元 B 更近, 即与 B 更相似, 这样, 我们就将 A 和 B 归为一类。

对一景影像的非监督分类过程就是通过成千上万次距离计算来判断像元之间，类别之间的相似性。通常，分析人员实际上并不知道非监督分类必须要计算的这些距离是多少，因为，计算机没有给出必要的中间步骤，仅仅给出影像分类的最终结果。可以说，距离计算是非监督分类的核心。

但是并非所有的距离计算都是基于欧氏距离。另一个简单的距离计算是 L_1 距离，即波段间差值的绝对值的和（Swain and Davis, 1978）。对于上例而言，L_1 距离为 73 = （8+12+30+23）。非监督分类距离计算的方法还有很多，很多比较复杂的方法是通过对距离缩放来提高像元分类的效率。

7.3.4 典型非监督分类算法

目前已经发展了很多用于多光谱影像分类的非监督分类算法，典型的有 K 均值和 ISODATA。下面对这两个算法进行概述。

1. K 均值

K 均值是影像分类中最常用的一种分类方法，首先需要指定聚类数量、初始聚类中心，然后将所有像元指定某聚类中心。这样就产生了一组粗略的聚类。之后，重新将所有像元根据最近的均值重新聚类，并重新计算聚类中心。这个过程重复进行，直到前后两次聚类的聚类中心不再发生变化为止。K 均值的执行过程如下：

(1) 选择一个聚类数量 C。这需要事先对影像进行观察以便于所确定的聚类数量尽可能反映影像数据的自然特征，一般要依据一定的聚类规则；

(2) 在光谱空间中选择聚类中心，记为 $m_c(c=1, 2, \cdots, C)$，这样就产生了初始聚类。理论上来说，此时聚类中心的指定是随机的，只是要求任何两个聚类中心不能重复。为了避免产生异常的聚类结果，一般最好指定初始的聚类中心在整个数据空间均匀分布，同时这样也有利于收敛。

(3) 计算像元到各聚类中心的距离（一般为欧氏距离），然后将其指定到与其最近的聚类中心。这样就产生了一组聚类结果。

(4) 根据(3)的聚类结果重新计算聚类中心，记为 $n_c(c=1, 2, \cdots, C)$。

(5) 如果 $n_c=m_c$，则聚类过程结束。否则将 n_c 指定为聚类中心，执行步骤(3)并重新聚类。

2. ISODATA 聚类

ISODATA 聚类方法与 K 均值相似，只是该聚类方法要么在迭代的过程中或迭代结果后对聚类结果进行检查。一般要检查聚类的数量或者是聚类的光谱形状。

在任何一个合适的阶段，该算法能对聚类进行检验：

(1) 类别的像元点太少，没有实际意义。比如，就像在最大似然分类中像元的统计分布很关键一样，如果某个类别的像元太少，很难产生可靠的均值或方差；

(2) 任何聚类之间如果相距太近而将其硬性地分为不同的类别显然是不合适的，因此需要将其进行合并。

对于(1)可以遵循如下原则：如果有 N 维光谱空间，那么只有当聚类类别的像元点数目在10N以上的聚类结果才有意义；对于(2)，何时对相邻的聚类类别进行合并，可以通过评估类别之间光谱的相似性。相似性的计算可以利用聚类类别的像元在其光谱空间的距离进行度量。

除了(1)和(2)，ISODATA还可以对聚类结果的光谱形状进行判断。如果需要，那么向某个方向延伸较长的类别将会一分为二。是否需要拆分，可以通过指定每个光谱波段的标准差数值，超过该数值的聚类结果将会被拆分。

7.3.5 非监督分类过程

一个典型的非监督分类过程一般首先需要制定分类类别的最小和最大数量。分类数量的确定主要是依据分析人员对最终的分类结果具有多少个类别的需求。分类算法首先确定一组像元作为聚类中心，聚类中心的确定一般是随机的，这样可以确保分析人员不会影响分类的结果，同时确保所选择的像元代表了整景影像中的数值。其后，分类算法计算像元之间的距离，并基于分析人员输入的限定条件产生对聚类中心的首次评估。一个类别可以用单独的一个点来表征，该点即为聚类中心，并将其作为某个给定类别的像元聚类中心，尽管很多分类过程并不总是将该中心作为该类的唯一中心。从这个意义上说，类别中仅包含经过首次评估后作为分类中心的像元。下一步，影像中所有剩余的像元被指定到从属于其与最近的分类中心的类别。这样，影像完成的分类，但是这一分类结果仅作为最终分类结果的一个参考，因为根据最初的聚类中心得到的分类结果可能并非是最优的分类结果，并不能满足分析人员所指定的限定条件。

下一步，因为随着新的像元的加入，意味着最初的分类中心不再准确，因此分类算法开始为每个类别寻找新的聚类中心。然后对整景影像再次进行分类，同时再次产生了新的聚类中心。如果该中心与该过程进行前寻找的中心存在差距，那么分类过程重复进行直到分类前后分类中心的位置没有明显变化，同时分类结果满足分析人员的要求为止。

在整个分类过程中，分析人员通常不对分类进行干涉，因此，非监督分类在分析人员限定条件下的被动分类。同时，非监督分类从影像中所鉴定出的自然结构是通过找到具有一致性的像元形成一个单一的类别，该过程没有受到分析人员已有的对像元属性或分布等知识的影响。然而，整个分类过程不能看成是完全被动的，因为分析人员已经通过对数据查看做出了一些决策，比如使用什么算法，分类的数量，甚至有可能对类别的特性和均一性也有一定的判断。这些决定会影响最终分类结果的性质和准确度，因此，非监督分类并非是没有考虑相关信息的分类。

目前可用的非监督分类过程有很多。虽然各有特色，但是大部分就是基于上述所描述的一般策略展开的。尽管，有些精炼后的分类过程可能会提高计算的速度和效率，但是在实质上依然是通过不断地将像元聚类来实现分类的。任何一个非监督分类算法的关键主要包括数据空间中有效的距离度量方法，聚类中心的确定以及类别独特性的测试。对于上述每个任务，都有很多策略可以实现，使用最广泛的方法这里就不一一列举了，感兴趣的同学可以查阅相关文献。

7.4 监督分类

对监督分类，非正式的定义为：利用事先采集的各类别的样本(即这些样本像元已经归为相应的信息类了)来将未知类别的像元归类(即将未分类的像元指定到某一信息类中)的过程。已知类别信息的样本是位于训练区的像元。训练区的确定主要是根据分析人员对影像中已知类别的范围有很清晰的认知后在影像中圈定的。这些区域要能够代表各所属类别的典型光谱特性，当然，必须与所对应的信息类具有很好的一致性。也就是说，训练区不能包括未知区域，当然也不能介于两个类别之间的边界上。无论是在影像上，还是在实地，尺寸、形状和位置都必须满足便于确认。从这些区域采集的像元就是训练样本，以用于引导分类算法将特定的光谱数值指定到适合的信息类中。很显然，这些训练数据的选取是监督分类的一个关键步骤。

7.4.1 监督分类的优点

相对于非监督分类，监督分类的优点如下：

(1)分类人员可以根据某一特定目的和地理区域来确定所要分的信息类的清单。如果要对比不同时期，同一地区的分类结果或者分类结果必须与邻近地区相协调时，这一特性尤为重要。在这些情况下，非监督分类所得到的不可预见的类别属性(比如类别数、类别特征、尺寸和模式等)可能是不适合或者不便利的。

(2)监督分类与已有认知的某一特定区域紧密相关的，通过选择训练区来达到分类的目的。

(3)采用监督分类方法，分析人员无需将最终的分类类别与其感兴趣的信息类别相匹配。

(4)通过将训练数据与分类结果进行对比，分类人员可以检测出分类过程中是否存在严重的错误，如果训练样本类别的准确度不够，说明要么分类过程要么样本选取过程存在严重的问题，尽管样本数据的正确分类并不总是意味着其他数据的分类也正确。

7.4.2 监督分类的缺点

相对于监督分类，监督分类有如下缺点：

(1)执行监督分类实际上是分析人员将某一分类结构强加给影像数据(也就是非监督分类所谓的"自然"类)。这些操作人员所定义的类别也许不能与数据中存在的自然类相匹配，因此在多维数据空间中并不是很清晰或者定义的很准确。

(2)训练数据的选取通常首先参考信息类别，其次才会考虑光谱属性。一个"100%林地"的训练区对于"森林"这一定义来说是正确的，但是在林地内还依然会有密度、年龄、阴影等诸如此类的差别，从这个意义上来说，这个训练区并不具有代表性。

(3)分析人员所选取的训练数据可能并不能反映整个影像的全部信息，尽管分析人员尽了最大的努力，但是这种情况还是存在的，尤其是当需要分类的面积很大、很复杂或者不能实地勘察时更易发生。

(4)即使手头有丰富的样本数据源,但是持续的训练数据的选取依然是一项耗时、花费高且枯燥的工作。分析人员在将地图上所确定的训练区与要分类的航拍影像匹配时可能会遇到问题。

(5)在训练数据中表现不出来的独特类别,那么监督分类可能无法识别,这可能是分析者事先未知或者这些类别在影像中仅占有非常小的区域。

7.4.3 训练数据

训练区指的是在数字影像中勾勒出来的已知类别的区域,一般通过在影像坐标中指定方形或矩形的角点和行列数来确定。当然,分析人员必须要已知每个区域的类别是什么。通常,分析人员需要首先收集和研究地图以及要分类的航片,并做实地考察。这里,我们假定分析人员对研究区的特定区域已经做了野外调研,对研究中存在的特定问题也比较熟悉,同时,在训练区的选取之间已经做了必要的野外勘察工作。为每一个信息类选区特定的训练区域,选取时遵循下列规则。主要目的是确定一组像元,它们能够正确地表征每个信息类区域的光谱变化(如图 7.5)。

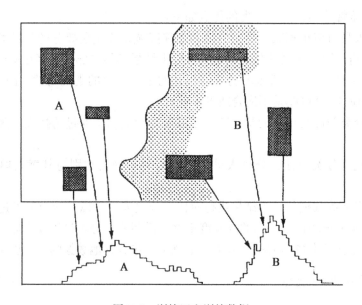

图 7.5 训练区和训练数据

每个训练区都包含很多像元,这些像元能够代表信息类的光谱特性。图中阴影区域代表训练场,每个训练场位置的确定都需谨慎,确保能够代表每个类别的光谱属性。这一信息为训练区以外其余像元的分类奠定了基础

1. 训练区的主要特性

在监督分类中,训练数据的选择是最困难也是最为关键的一部分。选择训练区的标准就是所选择的训练区应具有代表性,能够代表需要分类的每一类别的特征。

像元数

训练样本的选取,很重要的一点就是为每一类别所选择的像元数目。

Chen 等（2002）研究表明，每个类别选取 25 个像元明显不够，对于低分辨率遥感数据而言，50~100 个像元结果会更好些。Joyce（1978）参考 Landsat MMS 数据，提出就像元数目而言，每个训练区选择 10~40 个像元。当然，对于 TM 和 SPOT 数据而言，这一数字就要发生变化，因为这些传感器的分辨率与 MSS 有一定差别。此外，每个训练区的最优尺寸随每个场景内每个类别的差异性而变化，分析人员基于已有经验，根据具体情况确定训练区的大小。

一般而言，对于一个 N 维光谱空间，最小的训练样本数为 N+1。但是由于很难保证像元之间的独立性，因此训练样本的最小数量要比 N+1 多得多。在实际操作中，切实可行的方法是为每个类选择 10N 个像元数，如果可能 100N 个则更好。

形状

训练区的形状不重要，只要该形状不影响在影像上正确勾勒和定位就可以了。通常，为方便起见，人们倾向于使用方形或矩形形状的训练区。因为这样的形状需要指定的结点数最少，通常，结点的确定是分析人员最头痛的事情。

位置

位置很重要，因此，每个类别都应该在整个影像上寻找几个不同的训练样本区域。样本区的位置要保证能够比较准确和方便地将该区域的轮廓从地图和航空影像中转换为数字影像。因为，训练数据要能够反映影像中的变化，所以，训练区域不应该集中在影像中最能代表该类别的地方，而是要能够反映影像整体情况的多个训练区域进行选取。

对于分析人员而言，最好能够利用野外勘察获取的数据作为训练样本，但是训练样本选取的要求通常与实际情况的限制相互冲突，因为去遥远的或无人区来采集训练数据以形成好的训练区是不现实的。一般来说，航空观测，或者对已有地图和航片的充分利用对于在研究区精确地勾画出训练区是很有帮助的。尽管这些实践听起来很合理，但是如果有野外采集数据做参考的情况下，我们不宜不顾及野外数据而采用直接在影像上选取样本数据直接进行分类的方法。

数量

训练区域的最佳个数取决于我们要分类的数量、密度以及可用于训练样本选取的资源。理想上而言，每个信息类或光谱子类最少要选择 5~10 个训练区域来确保每个类别的光谱属性。Chen 等（2002）的研究表明，无论是低分辨率卫星数据还是高分影像，为每个类别选择 25 个训练区可以获取很好的分类精度。通常信息类在光谱上也是有差异的，因此有必要为每一个信息类选择几个训练区的数据，以将光谱子类的信息考虑进来。因为在后期的分类过程中，如果训练数据不合适的话，可能会抛弃一些训练区，因此选择多个训练区也是必要的。实践经验表明，最好选择多个小的训练区，而不是选择几个大的训练区。

布局

训练样本应该放置在能够方便且准确确定类别特征的地方，比如水体或影像中不同特征间具有明显边界的地方。训练区应该布设在整个影像中以便于它们能够代表类别在影像中的分布密度。训练区的边界应该远离具有不同性质地块的边缘，这样就可以避开边界处的像元。

均一性

训练区的均一性或许是训练样本选取中最重要的属性。在每个训练区域中，每个光谱波段的数据应该具有单一的频率分布（图 7.6）。如果训练区像元数值的直方图呈双峰分布，并且通过调整边界位置样本的均一性还是不够好的话，那么应该抛弃这样的训练区。训练数据可以用来计算不同光谱波段光谱数值的均值、方差和协方差。对于每一个类别，这些估计值是每个波段均值、波段差异性和波段关系的度量。最理想的是这些数值能够代表影像中每个类别，那么就可以用这些数值作为影像中位于训练区之外的大量像元分类的基础。当然，在实际应用中，影像很复杂，变化很大，且每个分析人员对该区域的认知不同，定义的训练区也不一样。此外，一些信息类在光谱上存在差异，单一的训练数据并不能较好地表达这样的信息类。

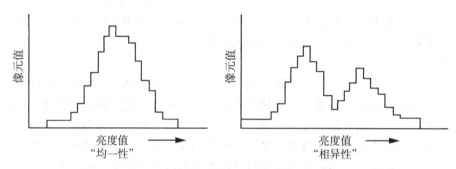

图 7.6 训练数据的均一性和相异性

左图中，训练数据的直方图呈单峰分布，表明光谱的均一程度。从这些训练区选取的样本数据很适合于作为影像分类的基础。右图中，另一组训练数据的直方图呈双峰分布，表明这一区域包含了两个光谱类。这样的训练区不能满足影像分类的需求，应该丢弃或者重新定义

2. 训练数据的重要性

Scholz 等（1979）和 Hixson 等（1980）对美国农业区分类精度的研究表明，训练数据选取的重要性绝不亚于分类算法的选取。他们断定训练数据的好坏对精度的影响比不同分类过程对精度的影响要大。如果使用相同的训练数据，几个不同分类算法所得到的分类精度差别不大。然而，对于同一个算法，如果两次训练数据不同则结果差异很大。这一发现表明，训练数据的选取，至少对某些分类实例来说，与分类器的选取同样重要。Scholz 等（1979）研究证明，训练区最重要的方面是所有的覆盖类型在每个光谱子类都应该采集到足够多的训练样本。

因为分类精度在很大程度上受限于训练数据的选取，所以 Campbell（1981）对训练数据的特性做了仔细的研究。其研究结果表明，不同训练区相邻的数据之间具有相似的数值。因此，在一个给定类比中，在一个训练区中采集的训练样本可能不是相互独立的。在相邻地块采集的训练样本可能会降低类内的差异，同时可提高类比的独特性。他的研究还显示出，两个土地覆盖类之间的相似程度随着波段和数据采集的时期的变化而变化。如果在一个类别中，训练样本的选择是随机的，而不是在各区域逐像元采集，那么这种高相似性就达到了最小化，进而提高了分类精度。他的研究结果也表明，用多个小面积的训练区

比用几个大的训练区其分类结果可能要好得多。

3. 训练数据选取的理想过程

执行监督分类的条件千差万别，因此详细的讨论训练数据选取的过程是不可能的，这在一定程度上受限于已有的设备和软件。然而，提出一个理想化的过程用作训练数据的选取和评估中的关键步骤还是可以的。

(1) 信息采集：包括需要测图区域的地图和航片。

(2) 开展实地研究：获取关于研究区的第一手信息。实地研究需要花费的时间因分析人员对研究区的熟悉程度而异。如果研究人员对研究区非常了解，且已经获得了最近的地图和航片，那就没有必要开展实地研究。但在大部分情况下，有些实地研究还是必须进行的。

(3) 详细计划野外考察用品，选择一个能够踏勘所有研究区的路线：地图是影像需要带到实地以便于分析人员能够边考察边注记（比如，影像可能要事先用透明纸等覆盖，纸质地图可以用彩色笔标记等）。尽管踏勘时间有限或者研究区进入困难，但要尽量保证能够观察到整个研究区内所有类别，不能仅局限于易于观察到的部分研究区。分析人员需要详细记录，关键部分认真注记在影像上。拍摄一些照片以作为对观察区的永久记录是非常有帮助的。在非常偏远的地区，航片可能是最可行的勘察研究区的方式了。最好的情况是野外踏勘时间与影像的获取时间一致，如果做不到二者一致，至少也要发生在同一个季节。

(4) 数字影像的初步检查：确定一些地标以便于确定训练区。评估影像质量，查看数据的频率直方图，分析是否有必要进行数据预处理。

(5) 圈定潜在的训练区：可参考 Joyce（1978）所给出的前述建议。训练区的大小需要根据地图或航片与数字影像的尺度差异而定。类别训练区的位置，类别特征无论在地图还是在影像上都应该易于辨别。

(6) 在数字影像上定位和勾勒训练区：确保将训练区圈定在同一地块之内以杜绝在训练区内出现混合像元。在这一阶段，所有的训练区都要按照影像的行列数来确定。

(7) 显示和检查每个训练区所有光谱波段的频率直方图。如果可能，要查看一下均值、方差、离散度、协方差等，以便于评估训练数据的有效性。

(8) 修改训练区的边界以消除双峰频率分布，或者，如果必要，丢弃具有双峰分布的不适合采集训练数据的训练区。在必要的情况下，可以转到步骤(1)重新定义新的训练区来取代这些不能用的训练区。

(9) 将训练样本的信息进行整合以便于利用其进行分类处理，下面将讲述分类过程中对这些数据的处理。

7.4.4 监督分类方法

监督分类的执行有很多种不同的方法。所有的分类都基于训练数据的信息作为训练区之外的像元的分类基础。目前已经发展了很多监督分类的算法。其中使用频率最高的有平行六面体、最小距离以及最大似然等分类器。

1. 平行六面体分类器

平行六面体分类器首先根据训练数据集计算每个类别 DN 值的变化范围，然后判断未知像元的 DN 值落入哪个类别的变化范围而将其指定为某个类别，如果该 DN 值不在任何类别的 DN 值变化范围中，则将其指定为"未知"类别。图 7.7 为平行六面体分类的示意图。

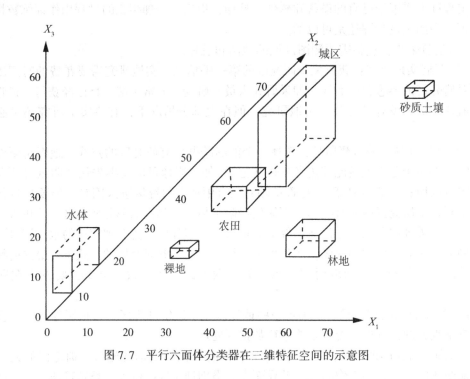

图 7.7 平行六面体分类器在三维特征空间的示意图

平行六面体分类可由表 7.1 中的数据说明。从 Landsat MSS 数据集中选择波段 5 和 7 的一部分数据即可以比较精确且易于理解地阐释平行六面体分类。图 7.8 中，X 轴是波段 5 的数值范围，波段 7 的最低值和最高值绘制在 Y 轴，然后将波段 7 和波段 5 的数值范围投影到 XOY 平面，由二者相交得到的多边形区域即定义了数据空间中各分类类别的范围。现在我们考虑一个未知类别的像元将其归类，其落入哪个由训练数据确定的类别区域就从属于哪个类别。如果必要，该过程可以扩展到许多波段，许多类别。此外，类别的边界最好通过训练数据的标准方差而不是数据的最大值和最小值。由于有个别像元会落入未分类类别中，因此，尽管这种策略比较有效，但也增加了在光谱数据空间中各类别相互叠加的几率。

尽管该分类器具有精度高、直接和简单的优点，但其缺点亦很显著。此分类器的主要问题在于两个或多个类别可能相互重叠。当发生重叠时，未知像元要么被归为"未知"，要么就简单地将其指定为其中之一。发生此问题可由如下原因造成：

（1）训练数据可能会低估类别的实际数值范围，因而在数据空间和影像中大面积的像元需要分类到各信息类别中。

（2）每一类类别中，像元并非均匀地分布在数据空间中，那些在类别边界附近的像元可能被划分到其他的类别中。

表 7.1　　　　　　　　　　　　　　图 7.8 中示例的数据

	数组 A				数组 B			
	波段				波段			
	4	5	6	7	4	5	6	7
	34	28	22	3	28	18	59	35
	36	35	24	6	28	21	57	34
	36	28	22	6	28	21	57	30
	36	31	23	5	28	14	59	35
	36	34	25	7	30	18	62	28
	36	31	21	6	30	18	62	38
	35	30	18	6	28	16	62	36
	36	33	24	2	30	22	59	37
	36	36	27	10	27	16	56	34
高	34	28	18	10	27	14	56	28
低	36	36	27	3	30	32	62	38

注：这些数据是从一个大的数据中选择出来的。

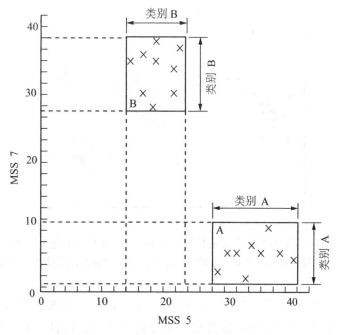

图 7.8　平行六面体分类中训练数据的类别范围(数据源于表 7.1)

这里只给出两个波段，但是基本原理可以扩展到多个光谱波段。训练区之外的像元如果落入训练数组组成的多边形中就从属于该类别此外，如果训练数据未能代表影像中数据的范围(这种情况发生的几率还很高)，那么就会造成影像上大面积像元归于未分类类别，或者需要修改这一基本过程来重新将这些像元指定到逻辑类别中。

2. 最小距离分类

最小距离分类器就是根据训练来计算各个类别的平均光谱信息。训练区的光谱数据可以在多维数据空间中绘制出来，就像前述的非监督分类的做法一样。由每个类别的训练数据所形成的聚类中，每个像元的位置是由其几个波段的数值决定的(图7.9)。这些聚类或许与前述所定义的非监督分类的结果相似，但是，在非监督分类中，这些聚类的像元是根据数据的自然结构来定义的。而对于监督分类来说，这些聚类是由分析人员所采集的训练区内的像元的数值决定的。

每个聚类可以利用其聚类中心作为代表，一般聚类中心是由数据的均值定义的。当需要将未分类的像元指定到某个类别中时，首先需要计算像元到每个聚类中心的多维距离，并将其指定到具有最小距离的聚类中。这样，分类过程总是根据一个像元到某个距离中心的最小距离来决定这些像元归于哪个类别。

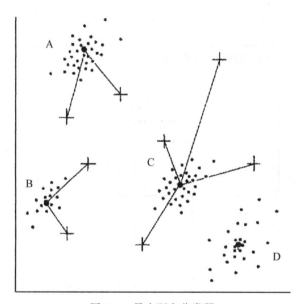

图7.9 最小距离分类器

图中的小点代表了训练数据所在的位置，十字线表示影像中其他地方未分类的像元，每个像元都被指定到与其最近的聚类中心的类别中，距离的计算是用前述的方法计算的

最小距离分类器不但易于理解，也易于执行，但是在遥感分类中并未得到广泛应用。这个分类器由于对光谱相应数据的方差变化不敏感，因此，对于在特征空间光谱类别相近或者光谱类别差异巨大时，最小距离均很难发挥作用。依据"用不同测距和不同方法所定义的聚类中心"这一基本途径来设计更复杂的最小距离分类方法，在理论上是可行的。

3. 最大似然分类

本质上，我们分类的类别展示了它们光谱模式的自然变化。此外，大气、地形造成的阴影、系统噪声、混合像元等都会对分类造成影响。因此，遥感影像很难单纯地记录光谱类别，更多记录的是逐波段的亮度值范围。到目前为止，我们所讨论的分类策略没有考虑到光谱类别内所呈现的变化，也没阐述当类别叠加后引起的光谱数据频率分布变化对分类的影响。比如，对于平行六面体分类来说，类别叠加是一个非常严重的问题，因为光谱空间不可能被分成相互独立的类别。这种情况经常发生，因为通常我们的注意力主要集中到光谱相近的像元，而不是分类器易于区分且能精确分类的像元。

因此，发生图7.10所描述的情形非常普遍。假设我们查看一景数字影像，其中包含四分之三的林地和四分之一的农田。林地和农田在平均亮度上差别很大，但是其极值（比如非常亮的林地像元或非常暗的农田像元）在两类地物重叠区的频率分布却很相似。（为了能够阐明这个问题，图7.10的数据仅仅是一个光谱波段的数据，当然原理是一样的，可以延伸到多个波段或者多个类别。）当亮度值为45时，就落入了重叠区，我们就不能断定这样的像元是输入林地还是农田。利用我们上述提到的决策规则，我们就不能断定哪个类别可以接收这样的像元，除非我们主观地划定类别的界限。在这种情况下，一个有效的分类就是考虑"45作为林地成员"和"45作为农田成员"的相对可能性。那么基于训练数据提供的参考信息，我们就选择一个具有最大正确性可能的类别。这种策略称为最大似然分类——利用训练数据估算各类别的均值和方差，根据均值和方差来估算可能性大小。最大似然分类，不仅考虑均值和分类像元的数值，还考虑了每个类别中亮度值的变化。

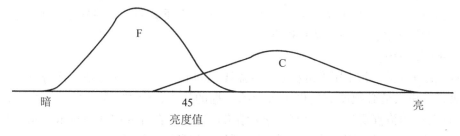

图7.10 最大似然分类

这是两个训练区像元的频率分布。重叠区域的像元可能属于F也可能属于C。依据每个类别中重叠区域像元值与总的像元值的关系，可以将像元指定到类别中

最大似然分类器假定每个类别每个波段的训练数据的统计信息均呈高斯正态分布。基于此，一个类别的光谱响应模式的分布可以通过均值和协方差来测量。要确定一个未知类别像元属于哪个类，该分类器需要计算这个像元属于各类别的概率值，然后将其指定到具有最大概率的类别中。当然，首先分析人员需要事先给定概率值的阈值，如果计算得到的概率小于给定的阈值，则该像元不归入任何一个类别。该分类需要计算机具有很好的计算性能，因此它比其他的简单分类技术需要更多的计算资源。同时，与其他监督分类技术相比，最大似然分类对训练数据的质量要求更高。对似然估计的前提是训练数据和类别数据都具有多方差的正态分布。（这就要求训练数据呈单峰分布）。遥感数据一般很难严格遵

循正态分布这一规则，但是由于一般接近正态分布，因此此分类用得很广。但是训练数据必须仔细选择，否则可能会带来误差。

7.4.5 执行监督分类的理想化次序

监督分类的执行较非监督分类的执行要复杂很多。分析人员必须要对每个分类步骤进行评估，如果不满意则返回到前一步，这样不断地精练和校正才能确保最终分类结果的正确。

(1) 确定需要分类的类别清单。这些类别最好能够与用户感兴趣的最终分类结果的类别相对应，但是实际上用户的需求可能比较模糊，或者用户需要分析人员的协助来确定一个合适的影像分类方案。

(2) 选择和定义训练数据。为每一类选择已知的、有代表性的像元。这些像元即为训练数据。训练数据可以来自于实地采集、地图、航空照片甚至是从影像产品上识别。由共同像元组成的边界所围成的区域即为训练区。

在这一步中需要对计算机上显示出来的影像与地图或者航片上所标记的训练区域进行仔细对比。之所以如此，是因为无论在影像上还是在地图上，训练区域都必须能够根据类别的独特特征分辨出来。然后，影像处理系统利用这些多边形边界的影像坐标来确定训练区中的像元以用于影像分类。所有的训练区都确定好后，分析人员就可以通过训练数据的频率分布来评估每类训练数据的均一性问题，并提取和评价每一类训练样本的参数以便选择适宜的分类算法。提取的参数称为各类别的标识信号或特征。

(3) 必要的情况下修改类别和训练区以便于定义更均一的训练数据。当分析人员查看训练数据时，或许需要从分类清单中删除、合并或者重新划分一些分类类别。如果经过修改后类别的训练数据达到了尺寸和均一性的要求，那么他们就可以用于分类；否则，必须返回到步骤(1)中。

(4) 执行分类。每个影像分析系统都需要不同的执行命令来完成一次分类，但是最核心的就是分析人员必须能够提供进入训练数据的程序(通常写在一个特定的计算机文件中)，同时必须检查要分类的影像。分类结果可以显示在屏幕或其他显示终端，并制作分类后影像的专题图及表格以便对各类别的面积等信息进行统计。

(5) 评估分类的性能。最后，分析人员必须对分类的结果做一下评估。如何评估，将在下一章进行讲述。

这一过程是执行监督分类的基本步骤。各步骤的细节随着分类方法和影像处理系统的不同而不同，但是其基本原理和涉及的问题是一致的，那就是精确地定义均一的训练区。

7.5 纹理分类

由于景观的光谱特性各异，因此影像分类是数字分析中一个比较独特的问题。理想化的地面景观区域包含地表光谱均一的地块。这些区域的精确成图可以通过将光谱类别和信息类的光谱"信号"相匹配来实现，就像上述所描述的分类一样。

在上述各分类器中，均没有考虑相邻像元的分类结果如何。在任何一景影像中，相邻像元是相关或相互影响的因为传感器所采集到的像元的信息会包含有邻近像元的信息，当然地面上某类地表覆盖类型的分布通常要大于一个像元的空间分辨率的大小。比如，低密度的居住区，当我们从上空俯视时，见到的是草地、街道、车道以及停车场等。在分类中，我们所感兴趣的是这些特征的融合而不是对各个组成部分的调查，因此各组成部分的调查本身对分类意义不大。因此，分类应该重点关注标识每个类别变化的总体模式而非平均亮度，因为平均亮度并不能表征类别之间的本质差异。尽管人工解译能够比较直观地识别这种复杂的模式，但是许多数字分类算法在对这些场景分类时却遇到了很多严重的问题，因为分类是将不同的光谱区域作为分散的信息类别来分类的。

当对一个像元进行分类时同时考虑与其邻近像元的可能类别，这种分类方法被称为是纹理敏感的分类方法，即纹理分类。可以实现纹理分类的途径很多，用得最多的是分类前影像处理，如中值滤波以及分类后空间滤波等方法。比如，统计一个具有某一特定尺寸的邻近区域亮度值的标准差，然后按一定顺序在整个影像上滑动，以标准差与像元数值的距离作为纹理标志，那么可以通过计算距离来检测出影像中光谱的变化。分析人员可以通过距离的大小来对各组成类别进行分类。

通常，复杂的计算纹理的方式可以产生更加令人满意的结果。比如，有的纹理测量是通过检测距离某一中心像元在不同距离和不同方向上亮度值的关系，并在整个影像上滑动来完成纹理的计算（Haralick 等，1973；Maurer，1974；Haralick，1979）。

Jensen（1979）研究发现，用 Landsat MSS 波段 5 数据，采用纹理进行分类，可以提高在郊区和城乡结合处分类的水平。分类精度提高，主要归功于在地物类别突变的边界处纹理信息变化较大的缘故。

当所定义的邻近区域相对较大时，比如 64×64 个像元时，纹理分类的效果较好。但是这样大的区域在遇到类别的边界处时也会出现问题。此外，这样大的区域也可能降低最终分类图有效的空间分辨率，因为纹理分类的结果是按照最小的空间元素来组织的。

7.6 模糊分类

模糊分类阐述的是一个暗含在很多事物处理中的问题，即每个像元都必须被指定到一个分散的类别中。尽管许多分类都尝试着最大程度地保证分类的正确性，但是基本的逻辑框架就是在像元和类别之间进行一对一的匹配。然而，我们知道，很多处理过程不能实现在像元和类别之间进行如此清晰的匹配。因此，将我们的关注点集中到寻找在像元和类别之间的匹配就使得许多像元的分类是错误的。模糊分类就是尝试着应用一个不同的分类逻辑来解决这一问题。

模糊逻辑已经应用到很多领域（Kosko and Isaka，1993），但是对于遥感来说则具有特殊的意义。模糊逻辑允许部分隶属，因为部分隶属与解决混合像元的问题很相近。所以这一特性对于遥感来说至关重要。比如说，传统的分类器将像元要么归类为林地要么归类为水体，但是模糊分类却允许一个像元属于水体的隶属度为 0.3，属于林地的隶属度为 0.7，即这个像元并没有确切地归类到某一类别中。隶属度取值一般从 0（即非隶属）到 1（完全隶属),位于中间的值则表明其隶属于一个或多个其他类别(表7.2)。

模糊分类根据隶属函数将隶属赋予像元(图7.11)。类别的隶属函数由数据和类别之间的一般关系或者定义的规则来确定。在遥感分类,隶属函数更多的是由根据某一特定场景的训练数据得到,隶属函数描述的是类别的隶属度和波段亮度值的关系(图7.11)。

表7.2　　　　　　　　　　模糊分类类别中的隶属度

类别	像元						
	A	B	C	D	E	F	G
水体	0.00	0.00	0.00	0.00	0.00	0.00	0.00
城市	0.00	0.01	0.00	0.00	0.00	0.00	0.85
交通	0.00	0.35	0.00	0.00	0.99	0.79	0.14
林地	0.07	0.00	0.78	0.98	0.00	0.00	0.00
草地	0.00	0.33	0.21	0.02	0.00	0.05	0.00
农田	0.92	0.30	0.00	0.00	0.00	0.15	0.00

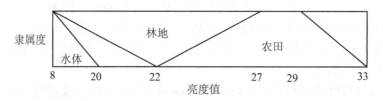

图7.11　模糊分类中的隶属函数

图7.11给出了几个像元和它们隶属度的示例(模糊分类的输出实际上是各像元隶属于某个类别的不同隶属度图像)。可以将像元隶属于某个类别的隶属度强硬的赋最高值1,而属于其他类别的隶属度为0,这与传统的分类一致,即一个像元一个类别,输出结果为所有像元都分类的影像。遥感图像处理程序可以调整隶属度,这样就可以调整类别结构和类别模式的连续性。

尽管对分类结果的评估很困难,但与传统的分类相比,模糊分类改进了分类的结果,至少在边缘上如此。但是对于主观的为像元赋予1隶属度的改进方法不予提倡,因为这样并未体现出模糊逻辑的功能。

本例中仅考虑单一波段数据分三个类别的隶属函数,当然可以扩展到多个波段。横轴表示亮度值,纵轴表示隶属度,从下到上表示由低到高。亮度值低于20的像元为水体,尽管低于8的亮度值是水体的可能性更大。农田的亮度值从22到33之间,尽管亮度值在27到29之间的像元才是农田。比如,亮度值为28的像元是农田,但亮度值为24的像元,部分属于林地,部分属于农田。未作标记的区域则不属于这三个类别。

表 7.3　　　　　　　　　　基于表 7.2 数据的强制分类示例

类别	像元						
	A	B	C	D	E	F	G
水体	0.00	0.00	0.00	0.00	0.00	0.00	0.00
城市	0.00	0.00	0.00	0.00	0.00	0.00	1.00
交通	0.00	1.00	0.00	0.00	1.00	1.00	0.00
林地	0.07	0.00	1.00	1.00	0.00	0.00	0.00
草地	0.00	0.00	0.00	0.00	0.00	0.00	0.00
农田	1.00	0.00	0.00	0.00	0.00	0.00	0.00

7.7　神经网络分类

人工神经网络(ANN)是通过建立和强化输入数据和输出数据之间的链路来模拟人脑学习过程的计算机程序。这些链路或路径可以模拟人脑的学习过程，因为在训练过程中输入与输出数值之间反复建立联系来强化链路，并以训练好的链路用于非训练区的数据的分类。

神经网络对数据的概率分布形式要求不高，它能够灵活地调整模型使得其能够满足系统的复杂性。因此，神经网络具备很强的妥协机制，这一点也是其引人关注之所在。

ANN 通常由三部分组成。由源数据组成的输入层，就遥感来说，源数据就是多光谱观测数据，可能是不同波段或不同日期的数据。ANN 需要输入的数据量非常大，包括多个波段多个日期的多光谱观测数据，以及相关的辅助数据。分析人员所需要的类别数据为输出层。尽管当输出类别数较少时，或者相对于训练数据的量来说，输出数据不是很多时，ANN 过程更可靠，但是我们几乎不能对输出层的特性做出限定。输入层的训练数据经过反复训练后在输入和输出之间建立一个明确的链路。在训练阶段，ANN 在一个或多个隐含层中通过调整输入和输出的权重来建立输入和输出的稳定联系(图 7.12)。在遥感中，通过训练数据，在分类类别和影像数值之间的反复训练来建立数据和类别的关系，可以强化隐含层的权重以便于 ANN 能够在非训练区对给定的光谱数据标识正确的类别标签。

此外，ANN 也可以通过后向传播(BP)网络来训练。如果将传统的影像分类训练形式成为前向传播的话，那么 BP 可以看做是对输入和输出链路的反向消除，即利用期望的输出结果和实际的输出结果之间的差异来调整权重。在此过程中，可以建立描述输入层和输出层之间关系的变换函数，以此来设置权重以强化在输入和输出之间的有效链接。比如，通过权重可以识别出有些波段组合对某些分类有效，有些对其他类别有效。在 BP 中，隐含层能够记录将数据匹配到各分类错误的同时调整权重以使错误达到最小。

相对于其他分类器，比如最大似然分类来说，ANN 使用的统计较少，尽管在实际中 ANN 的成功使用需要更加慎重。实践证明，ANN 在遥感数据分类中可以比较准确，尽管在精度上的改进较小或者一般。ANN 需要对训练数据做大量工作，因此需要具有更强计算能力的计算机。

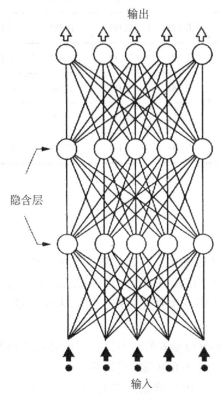

图 7.12　人工神经网络

7.8　数字影像分类的后处理

基于如下几方面的考虑进行数字影像的后处理：(1)需要手工编辑分类影像，比如陡峭的斜坡被分类为水体等，需要手工更正过来；(2)利用辅助数据细化影像分类结果；(3)平滑影像分类结果；(4)将分类数据转换为矢量格式，因为用户发现矢量格式比较容易输出，且很多数据都已经是矢量格式。

7.8.1　矢量与栅格数据的整合——由像元到多边形

因为分类后的影像存在太多的碎屑多边形，所以由像元到多边形的自动转换行不通，可以采用滤波的方法将像元聚集为较均一的类，然后将聚合的类自动转换为多边形。

7.8.2 空间滤波

空间滤波是对区域的操作。利用周围临近像元的数值，基于某种函数关系对像元重新计算数值。通常利用移动窗或核来平滑现有的分类图。主要的核操作包括：最大值、最小值、众数、寡数、均值、中位数、标准差以及偏差，等等。

7.8.3 移动窗

移动窗技术扫描每个像元周围临近 3×3 个像元的值(图 7.13)。通过核的类型，即某种函数类型，计算出该像元一个新值，扫描完成后形成新的影像图(图 7.14)

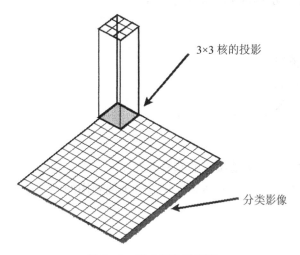

图 7.13　移动窗的示意图

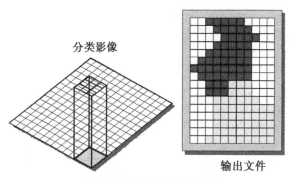

图 7.14　移动窗输出文件

这种通过移动窗实现空间滤波的图像后处理方法，可以明显地削弱基于像元分类的分类器所制作的专题图中的"椒盐现象"。但是，与此同时，这种方法也会将某些类别夸大以至于提高了错分误差(这部分内容下一章介绍)。

☞ **小结**

- 利用光谱或辐射特性的差异来区分地物
- 土地覆盖不等同于土地利用
- 监督分类

（1）利用训练数据来确定类别的光谱特性。

（2）将其他像元依据与训练样本光谱特性相匹配的程度进行分类。

- 非监督分类

（1）最大程度地将各分类结果分离。

（2）分类后对每个分类类别标识类别名称。

☞ **知识点**

- 分类的基本类型
- 训练样本的选取
- 分类后处理方法

☞ **思考题**

（1）请调查土地利用与土地覆盖的基本含义。

（2）思考从遥感影像上获取的信息反映的是土地利用的信息还是土地覆盖的信息，并解释原因。

第8章 精度评定

分类后还需要告诉使用分类结果的人，该分类结果与实际情况的符合程度。没有精度评定，分类结果仅仅是一张漂亮的图片。精度评定有助于我们评估分类结果的好坏，同时利用精度评估也可以解释分类结果的有用程度。

8.1 参考数据

8.1.1 参考数据来源

参考数据来源主要包括：解译的航片，具有 GPS 控制点的地面真实数据以及已有的 GIS 图。

8.1.2 参考数据源的选择

确保能够从参考数据源中提取出分类方案所需要的信息，比如，如果分类方案是分出四种不同种类的草，那么航片是不能满足要求的。最好需要有 GPS 点位的地面数据。

8.1.3 确定参考图的大小

参考图与遥感数据在空间尺度上要相匹配，也就是说，如果遥感数据的空间分辨率为 1 km，那么具有 5 m 分辨率的 GPS 点位数据就没有意义了。这种情况下，航片或其他的卫星影像也许更合适。

考虑影像的空间频率，比如图 8.1 的两个例子，参考数据能够在某一方向上覆盖影像的 3 个像元，对于不同的空间频率，结果明显不同。同时，还要考虑参考数据与分类影像的匹配程度，比如与图 8.2 同样的两张图，点位匹配误差差一个像元，结果差异很大。

8.1.4 确定参考数据样本的位置和数量

确定能够采集到足够的参考数据样本。参考数据的采集方法有很多，比如随机的、分层随机的和系统的方法，等等。

参考数据越好，统计评估结果越好。通常而言，参考数据越多越好，Lillesand 和 Kiefer 建议，经验上每个类别需要选 50 个像元。

选择了参考数据源，参考图的大小以及位置，接下来需要确定参考数据的类别、分类数据的类别，并将二者进行比较。

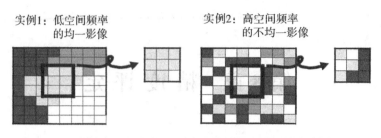

图 8.1 不同空间频率与参考数据的误差

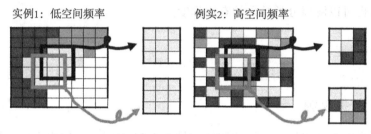

图 8.2 不同空间频率与参考数据的位置匹配误差

8.1.5 实例

表 8.1　　　　　　　　参考数据和分类影像图的比较

参考数据图号	参考数据源的类别	分类影像图的类别	一致性
1	针叶树	针叶树	Yes
2	阔叶树	针叶树	No
3	水体	水体	Yes
4	阔叶树	阔叶树	Yes
5	草地	阔叶树	No

8.2 精度评定方法

利用误差矩阵进行评估（表 8.2）。

表 8.2　　　　　　　　　　　　　　误差矩阵

	参考数据源的分类类别				
分类影像图的 分类类别	类别	针叶树	阔叶树	水体	总计
	针叶树	50	5	2	57
	阔叶树	14	13	0	27
	水体	3	5	8	16
	总计	67	23	10	100

8.2.1　精度评定之总体精度

总体精度：正确的点位数/总的点位数（表8.3）。

表 8.3　　　　　　　　　根据误差矩阵计算总体分类精度

	参考数据源的分类类别				
分类影像图的 分类类别	类别	针叶树	阔叶树	水体	总计
	针叶树	50	5	2	57
	阔叶树	14	13	0	27
	水体	3	5	8	16
	总计	67	23	10	100

$$总体分类精度 = \frac{50+13+8}{100} \times 100\% = 71\%$$

位于对角线格网内的数据表征的是与参考数据相比分类结果正确的数据，对角线之外的为分类错误的数据。

总体精度存在的问题：总体精度是一个平均值，它不能揭示误差是均匀地分布在各个类中还是有些类的分类精度很好而有些类的分类精度则很差。因此，还有其他形式的精度评估方法，包括用户精度和制图精度。

用户和制图精度以及误差类型：

用户精度与错分误差相对应，比如，1棵灌木和3棵针叶树被错误地分在草地类别中。

制图精度与漏分误差相对应，比如，7棵针叶树和1棵灌木被从草地中漏分。

8.2.2　精度评定之用户精度

从分类图用户的角度来评估分类图的精度如何（表8.4）。

对于一个给定的类别，有多少像元正确地落在它们应该在的地方。

用户精度计算为：给定分类图中被鉴定为正确的像元数或分类图中该类别的像元数。

表 8.4　　　　　　　　　　　　用户精度计算

		参考数据源的分类类别			
	类别	针叶树	阔叶树	水体	总计
分类影像图的分类类别	针叶树	50	5	2	57
	阔叶树	14	13	0	27
	水体	3	5	8	16
	总计	67	23	10	100

以针叶树为例：

$$针叶树的用户精度 = \frac{50}{70} \times 100\% = 88\%$$

8.2.3　精度评定之制图精度

制图精度：从分类影像图制作的角度来评估制图的精度如何。对于在参考数据的给定的类别，有多少像元在影像图上正确的标识出来。

计算为：给定类别参考数据上标识正确的像元数或参考数据上该类实际的像元数（表8.5）。

表 8.5　　　　　　　　　　　　制图精度的计算

		参考数据源的分类类别			
	类别	针叶树	阔叶树	水体	总计
分类影像图的分类类别	针叶树	50	5	2	57
	阔叶树	14	13	0	27
	水体	3	5	8	16
	总计	67	23	10	100

以针叶树为例：

$$针叶树的制图精度 = \frac{50}{67} \times 100\% = 75\%$$

上述精度评估结果(表8.6)。

表 8.6　　　　　　　　　　　　精度评估汇总表

		参考数据源的分类类别				用户精度
	类别	针叶树	阔叶树	水体	总计	
分类影像图的分类类别	针叶树	50	5	2	57	88%
	阔叶树	14	13	0	27	48%
	水体	3	5	8	16	50%
	总计	67	23	10	100	
制图精度		75%	57%	80%		总计 71%

8.2.4　精度评定之 Kappa 系数

1. Kappa 系数

Kappa 系数反映了实际的一致性与所期待的一致性之间的差异。比如 Kappa 系数为 0.85 意味着有 85% 实际的一致性与所期待的一致性相同。

Kappa 系数计算为：

$$\hat{K} = \frac{观测精度 - 偶然的一致性}{1 - 偶然的一致性}$$

观测精度是由误差矩阵中对角线格网中的数据决定的。

偶然的一致性是由非对角线的网格数值决定的，计算为：每一类中行列乘积的和。

2. Kappa 系数计算实例(表 8.6)

观测精度：71/100 = 0.71

偶然因素造成的一致性：

(57×67+27×23+16×10)/(100^2) = 0.4606

$$\hat{K} = \frac{0.71 - 0.4606}{1 - 0.4606} = 0.46$$

8.2.5　精度评定之评述

不同的精度评定方法产生不同的信息。如果我们只注意到其中一个，那么或许我们得到的精度评估信息是错误的。比如，在表 8.6 中，总体精度为 71%，但是阔叶树的用户精度却只有 48%。

1. 如何展示精度评定结果

(1) 取决于观众。

(2) 取决于研究目的。

(3) 大部分文献建议将误差矩阵、用户精度、制图精度以及总体精度、Kappa 系数等分别列出。

2. 精度偏低的原因

参考数据的错误：

(1) 位置错误——最好利用影像进行校正。
(2) 解译错误。
(3) 参考媒介与分类不一致。

分类影像图的错误，主要是遥感数据不能确定具体的功能类别：
(1) 分类是土地利用，而非遥感所确定的覆盖类别。
(2) 分类的地物在光谱上不具备可分离性。
(3) 大气效应可能会掩盖类别之间细微的差异。
(4) 传感器的空间尺度与分类方案中的分类类别不匹配。

3. 改进精度评估的方法

土地利用/土地覆盖可以与其他数据结合起来，如与高程、温度、产权、距离，以及纹理等信息相结合。

对于光谱的不可分割性，可以增加光谱数据，比如高光谱或者多时相数据。

大气效应可以采用大气校正方法。

空间尺度上可以改变光谱数据，改变传感器、像元聚合等。

在分类影像中的错误，可以采用高光谱的分类方案或者在最大似然分类中利用优先概率来提取出影像中较少的类别。

☞ 小结

- 参考数据的选取很关键，要考虑所用的传感器以及分类方案
- 误差矩阵是分类精度评估的基础
- 所有形式的精度评估结果都应该展示给用户
- 对精度的解释可以更好地改进分类的精度

☞ 主要知识点

- 参考数据的选取
- 精度评估的方法和指标

☞ 思考题

(1) 如果我们不能够有效地评定分类影像的精度，其后果会是怎样的？
(2) 请列举精度评定技术的可能应用领域。

Chapter 1 Overview of Remote Sensing

1.1 Definitions

The field of remote sensing has been defined many times as follows:

Remote sensing has been variously defined, but basically it is the art or science of telling something about an object without touching it. (Fisher et al., 1976, p. 34)

Remote sensing is the acquisition of physical data of an object without touch or contact. (Lintz and Simonett, 1976, p. 1)

Imagery is acquired with a sensor other than (or in addition to) a conventional camera through which a scene is recorded, such as by electronic scanning, using radiations outside the normal visual range of the film and camera—microwave, radar, thermal, infrared, ultraviolet, as well as multispectral, special techniques are applied to process and interpret remote sensing imagery for the purpose of producing conventional maps, thematic maps, resources surveys, etc., in the fields of agriculture, archaeology, forestry, geography, geology, and others. (American Society of Photogrammetry)

Remote sensing is the observation of a target by a device separated from it by some distance. (Barrett and Curtis, 1976, p. 3)

The term "remote sensing", in its broadest sense merely means "reconnaissance at a distance." (Colwell, 1966, p. 71)

Remote sensing, though not precisely defined, includes all methods of obtaining pictures or other forms of electromagnetic records of the Earth's surface from a distance, and the treatment and processing of the picture data... Remote sensing then in the widest sense is concerned with detecting and recording electromagnetic radiation from the target areas in the field of view of the sensor instrument. This radiation may have originated directly from separate components of the target area; it may be solar energy reflected from them; or it may be reflections of energy transmitted to the target area from the sensor itself. (White, 1977, pp. 1-2)

"Remote sensing" is the term currently used by a number of scientists for the study of remote objects (Earth, lunar, and planetary surfaces and atmospheres, stellar and galactic phenomena, etc.) from great distances. Broadly defined, remote sensing denotes the joint effects of employing modern sensors, data-processing equipment, information theory and processing methodology, communications theory and devices, space and airborne vehicles, and large-systems theory and practice for the purposes of carrying out aerial or space surveys of the Earth's surface. (National Academy of Sciences, 1970, p. 1)

Remote sensing is the science of deriving information about an object from measurements made at a distance from the object, i. e., without actually coming in contact with it. The quantity most frequently measured in present-day remote sensing systems is the electromagnetic energy emanating from objects of interest, and although there are other possibilities (e. g., seismic waves, sonic waves, and gravitational force), our attention focus upon systems which measure electromagnetic energy. (D. A. Landgrebe, quoted in Swain and Davis, 1978, p. 1)

Examination of the common elements in these varied definitions permits identification of the topic's most important themes. From a cursory look at these definitions, it is easy to identify a central concept: the gathering of information on distance. This excessively broad definition, however, must be refined if it is to guide us in studying a body of knowledge that can be approached in a single course of study.

The kind of remote sensing to be discussed here is devoted to the observation of the Earth's land and water surfaces by means of reflected or emitted electromagnetic energy. This more focused definition excludes applications that could be reasonably included in broader definitions, such as sensing the Earth's magnetic field or atmosphere, or the temperature of the human body. Also, we will focus upon instruments that present information in an image format, so we must largely exclude instruments (e. g., certain lasers) that collect data at a distance, but do not portray the result in image format. Such excluded applications can, of course, be considered "remote sensing" in its broad meaning, but they are omitted here as a matter of convenience. For our purposes, the definition is:

"Remote sensing (RS) is the science (and to some extent, art) of acquiring information about the Earth's surface by aircraft or satellite sensors without actually being in contact with it. This is done by sensing and recording reflected or emitted energy and processing, analysing, and applying that information."

1.2 Milestones in the History of Remote Sensing

The scope of the field of remote sensing can be elaborated by examining its history to trace the development of some of its central concepts. A few key events can be offered to trace the evolution of the field (Table 1.1).

Table 1.1 **Milestones in the History of RS**

1800	Discovery of infrared by Sir William Herschel
1839	Beginning of practice of photography

续表

1847	Infrared spectrum shown by A. H. L. Fizeau and J. B. L. Foucault to share properties with visible light
1850-1860	Photography from balloons
1873	The theory of electromagnetic energy developed by James Clerk Maxwell
1909	Photography from airplanes
1914-1918	World War I: aerial reconnaissance
1920-1930	Development and initial applications of Aerial Photography and Photogrammetry
1929-1939	Economic depression generates environmental crises that lead to governmental applications of aerial photography
1930-1940	Development of radars in Germany, United States, and United Kingdom
1939-1945	World War II: applications of invisible portions of electromagnetic spectrum; training of persons in acquisition and interpretation of air photos
1950-1960	Military research and development
1956	Colwell's research on plant disease detection with infrared photography
1960-1970	First use of term "remote sensing" TIROS weather satellite Skylab remote sensing observations from space
1972	Launch of Landsat 1
1970-1980	Rapid advances in digital image processing
1980-1990	Landsat 4: new generation of Landsat sensors
1986	SPOT French Earth observation satellite
1980s	Development of hyperspectral sensors
1990s	Global remote sensing systems

Evelyn Pruit, a scientist working for the U. S. Navy's office of Naval Research, coined the term "remote sensing" when she recognized that the term "aerial photography" no longer accurately described the many forms of imagery collected using radiation outside the visible region of the spectrum. Early in the 1960s the U. S. National Aeronautics and Space Administration (NASA) established a research program in remote sensing—a program that, during the next decade, was to support remote sensing research at institutions throughout the United States. During this same period, a committee of the U. S. National Academy of Science (NAS) studied opportunities for the application of remote sensing in the field of agriculture and forestry. In 1970, the NAS reported the results of their work in a document that outlined many of the opportunities offered by this emerging field of inquiry.

In 1972, the launch of Landsat 1, the first of many earth-orbiting satellites designed for observation of the Earth's land areas, formed another milestone. Landsat provided, for the first time, systematic and repetitive observation of the Earth's land areas. Each Landsat image depicted large areas of the Earth's surface in several regions of the electromagnetic spectrum, yet provided modest levels of detail sufficient for practical applications in many fields. Landsat's full significance may not be fully appreciated, yet it is possible to recognize three of its most important contributions. First, the routine availability of multispectral data for large regions of the Earth's surface greatly expanded the number of people who acquired experience and interest in analysis of multispectral data. Landsat's second contribution was to create an incentive for the rapid and broad expansion of the uses of digital analyses for remote sensing. The third contribution of the Landsat program was its role as a model for development of other land observation satellites designed and operated by diverse organizations throughout the world.

1.3 Remote Sensing Process

1.3.1 Overview of Remote Sensing Process

Because remotely sensed images are formed by many interrelated processes, an isolated focus on any single component will produce a fragmented understanding. Therefore, our initial view of the field can benefit from a broad perspective that identifies the kinds of knowledge required for the practice of remote sensing (Figure 1.1).

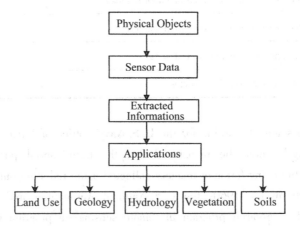

Figure 1.1　Schematic overview of RS process

Consider first the *physical objects*, consisting of buildings, vegetation, soil, water, and the like. There are the objects that applications scientists wish to examine. Knowledge of the physical objects resides within specific disciplines, such as geology, forestry, soil science, geography, and urban planning.

Sensor datas are formed as an instrument (e.g., a camera or radar) views the physical objects by recording electromagnetic radiation emitted or reflected from the landscape. For many of us, sensor data often seem to be abstract and foreign because of their unfamiliar overhead perspective, unusual resolutions, and use of spectral regions outside the visible spectrum. As a result, effective use of sensor data requires analysis and interpretation to convert data into information that can be used to address practical problems, such as siting landfills or searching for mineral deposits. These interpretations create extracted information, which consist of transformations of sensor data designed to reveal specific kinds of information. Actually, the same sensor data can be examined from an alternative perspective to yield different interpretations. Therefore, a single image can be interpreted to provide information about soils, land use, or hydrology, for example, depending on the specific image and the purpose of the analysis. Generally speaking, the information that can be remotely sensed listed below:

(1) Planimetric (x, y) location and dimensions.

(2) Topographic (z) location.

(3) Colour (spectral reflectance).

(4) Surface temperature.

(5) Texture.

(6) Surface roughness.

(7) Moisture content.

(8) Vegetation biomass.

Finally, we proceed to the applications, in which the analysed remote sensing data can be combined with other data to address a specific practical problem, such as land-use planning, mineral exploration, or water-quality mapping.

1.3.2 A Specific Example in Remote Sensing Process

Let's consider how a remotely sensed image is acquired by tracing the path of energy used to make an image (Figure 1.2).

(1) Incoming solar radiation is in part scattered or absorbed by the Earth's atmosphere.

(2) The remaining energy reaches the Earth's surface to interact with the features present on the landscape—in this example, a living tree.

(3) Much of the reflection of energy from the tree canopy is controlled by interactions with individual leaves, which selectively absorb, transmit, and reflect energy depending upon its wavelength.

(4) The reflected energy is again subject to atmospheric attenuation before it is recorded by an airborne or satellite sensor.

(5) The results of a multitude of such interactions are recorded as a photograph-like image,

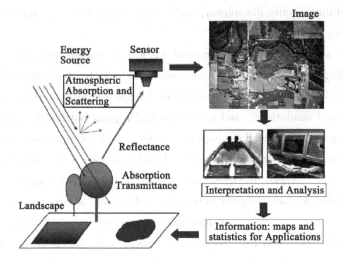

Figure1.2 A specific example of RS Process

or as an array of quantitative values.

(6) The image is then available for interpretation and analysis to derive information concerning the landscape.

1.4 Key Concepts of Remote Sensing

The practice of remote sensing is young enough that the basic facts and methods are not yet known completely. Scientists are still investigating to define many of the fundamental methods and concepts central to remote sensing. Nonetheless, it is useful to base our study of remote sensing upon a set of principles that seem to convey the essential dimensions of the practice of remote sensing.

1.4.1 Spatial Resolution

For some remote sensing instruments, the distance between the target being imaged and the platform, plays a large role in determining the detail of information obtained and the total area imaged by the sensor. Sensors on-board platforms far away from their targets, typically view a larger area, but cannot provide great detail. Compare what an astronaut on-board the space shuttle sees of the Earth to what you can see from an airplane. The astronaut might see your whole province or country in one glance, but couldn't distinguish individual houses. Flying over a city or town, you would be able to see individual buildings and cars, but you would be viewing a much smaller area than the astronaut.

The detail discernible in an image is dependent on the spatial resolution of the sensor and refers to the size of the smallest possible feature that can be detected. Spatial resolution of passive sensors depends primarily on their Instantaneous Field of View (IFOV) (Figure 1.3). The IFOV is the angular cone of visibility of the sensor and determines the area on the Earth's surface which is "seen" from a given altitude at one particular moment in time. This area on the ground is called the resolution cell and determines a sensor's maximum spatial resolution.

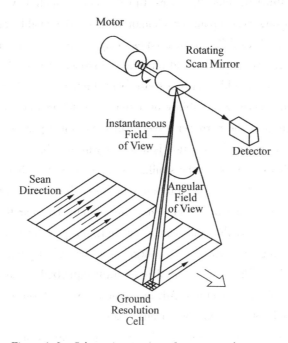

Figure 1.3 Schematic overview of a cross-track scanner

Images where only large features are visible are said to have coarse or low resolution. In fine or high resolution images, small objects can be detected. Military sensors, for example, are designed to view as much detail as possible, and therefore have very fine resolution. Commercial satellites provide imagery with resolutions varying from a few meters to several kilometers. Generally speaking, the finer the resolution, the less ground area can be seen.

The ratio of distance on an image or map, to actual ground distance is referred to as scale. If you had a map with a scale of 1 : 100,000, an object of 1 cm length on the map would actually be 100,000 cm (1 km) long an object lying on the ground. Maps or images with small "map-to-ground ratios" are referred to as small scale (e.g., 1 : 100,000), and those with larger ratios (e.g., 1 : 5,000) are called large scale.

Spatial resolution may also be described as the ground surface area that forms one pixel in the satellite image. The IFOV or ground resolution of the Landsat Thematic Mapper (TM) sensor, for example, is 30 m. The ground resolution of weather satellite sensors has been often larger than a square kilometer.

There are satellites that collect data at less than one meter ground resolution, but these are classified military satellites or very expensive commercial systems.

1.4.2 Spectral Resolution

Different classes of features and details in an image can often be distinguished by comparing their responses over distinct wavelength ranges. Broad classes, such as water and vegetation, can usually be separated using very broad wavelength ranges—the visible and near infrared. Other more specific classes, such as different rock types, may not be easily distinguishable using either of these broad wavelength ranges and would require comparison at a much finer wavelength ranges to separate them. Thus, we would require a sensor with higher spectral resolution.

Spectral resolution describes the ability of a sensor to define fine wavelength intervals. The finer the spectral resolution, the narrower wavelength range for a particular channel or band.

The simplest form of spectral resolution is a sensor with one band only, which senses visible light. An image from this sensor would be similar in appearance to a black and white photograph from an aircraft. A sensor with three spectral bands in the visible region of the EM spectrum would collect similar information to that of the human vision system. The Landsat TM sensor has seven spectral bands located in the visible and near to mid infrared parts of the spectrum.

Many remote sensing systems record energy over several separate wavelengths range at various spectral resolutions. These are referred to as **multi-spectral** sensors and will be described in some detail in the following sections. Advanced multi-spectral sensors called **hyperspectral** sensors, detect hundreds of very narrow spectral bands throughout the visible, near-infrared, and mid-infrared portions of the electromagnetic spectrum (Figure 1.4). Their very high spectral resolution facilitates fine discrimination between different targets based on their spectral response in each of the narrow bands.

1.4.3 Radiometric Resolution

While the arrangement of pixels describes the spatial structure of an image, the radiometric characteristics describe the actual information contained in an image. Every time an image is acquired on film or by a sensor, its sensitivity to the magnitude of the electromagnetic energy determines the radiometric resolution. The radiometric resolution of an imaging system describes its ability to discriminate very slight differences in energy. The finer the radiometric resolution of a sensor, the more sensitive it is to detect small differences in reflected or emitted energy. Imagery data are represented by positive digital numbers which vary from 0 to (one less than) a selected power of 2. This range corresponds to the number of bits used for coding numbers in binary format. Each bit records an exponent of power 2 (e.g., 1 bit=2^1=2). Thus, if a sensor used 8 bits to record the data, there would be 2^8 = 256 digital values available, ranging from 0 to 255. However, if only 4 bits were used, then only 2^4 = 16 values ranging from 0 to 15 would be available. Thus, the radiometric resolution would be much less.

By comparing a 2-bit image with a 6-bit image, we can see that there is a large difference in

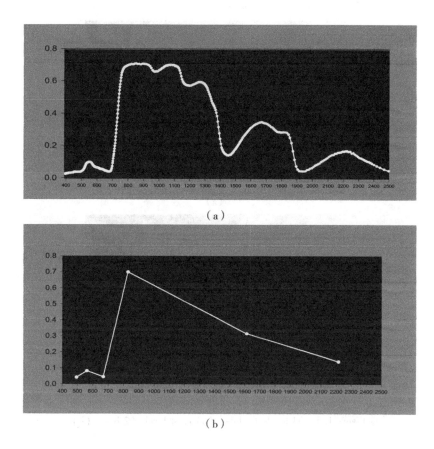

Figure 1.4 High spectral resolution (a) and low spectral resolution (b)

the level of detail discernible depending on their radiometric resolutions (Figure 1.5).

Radiometric Resolution indicates how finely does the satellite divide up the radiance it receives in each band? This usually expressed as the number of bits used to store the maximum radiance. For example, 8 bits would be the range of brightness values to 256 (usually 0 to 255).

- 1 bit (0~ 1).
- 8 bits (0~ 255): Everything will be scaled from 0 to 255, 0: No EMR or below some minimum value (threshold), 255: Max EMR or above some threshold for 8 bits data type. Subtle details may not be represented.
- 16 bits (0~ 65,535): Wide range of choices, Required storage space will be twice that of 8 bits.
- 32 bits (0~ 4,294,967,295) & more.

1.4.4 Temporal Resolution

In addition to spatial, spectral, and radiometric resolution, the concept of temporal

Figure 1.5 64 levels (6 bits, (a)), 4 levels (2 bits, (b))

resolution is also important to consider in a remote sensing system. The revisit period (i. e. temporal resolution) of a satellite sensor is usually several days.

The temporal resolution is the time lag between two subsequent data acquisitions for an area. For example: Aerial photos in 1971, 1981, 1991, and 2001, the temporal resolution is 10 years. The temporal resolution indicates how frequently a satellite view the same place. This depends on:

(1) Orbital characteristics.

(2) Swath width.

(3) Ability to point to the recording instrument.

The frequency will vary from several times per day for a typical weather satellite, to 8-20 times a year for a moderate ground resolution satellite, such as Landsat TM. The frequency characteristics will be determined by the design of the satellite sensor and its orbit pattern.

1.4.5 Geometric Transformation

Every remotely sensed image represents a landscape in a specific geometric relationship determined by the design of the remote sensing instrument, specific operating conditions, terrain relief, and other factors. The ideal remote sensing instrument would be able to create an image with accurate, consistent geometric relationships between points on the ground and their corresponding representations on the image. Such an image could form the basis for accurate measurements of areas and distances. In reality, of course, each image includes positional errors caused by the perspective of the sensor optics, the motion of scanning optics, terrain relief, and earth curvature. Each source of error can vary in significance in specific instances, but the result is that geometric errors are inherent, not accidental, characteristics of remotely sensed images. In some instances, we may be able to remove or reduce location error, but it must always be taken into account before images are used as the basis for measurements of areas and distances.

1.4.6 Interchangeability of Pictorial and Digital Formats

Most remote sensing systems generate photograph-like images of the Earth's surface. Any such image can be represented in digital form by systematically subdividing the image into tiny areas of equal size and shape, then representing the brightness of these areas by discrete values. Conversely, many remote sensing systems generate, as their first generation output, digital arrays that represent brightness of areas of the Earth's surface in digital form. Digital images can be displayed as pictorial images by displaying each digital value as a brightness level scaled to the magnitude of the value.

The two forms of remote sensing data—pictorial and digital—represent different methods of display and representation, but there is no real difference in the information conveyed by the two forms. Any images can be portrayed in either form (sometimes with a loss of detail in converting from one form to another) according to the purposes of our investigation.

1.4.7 Remote Sensing Instrumentation Acts as a System

The image analyst must always be conscious of the fact that the many components of the remote sensing process act as a system, and therefore cannot be isolated from one another. For example, upgrading the quality of a camera lens makes little sense unless we also use a film of sufficient quality to record the improvements produced by the superior lens, thus allowing the analyst to have the ability to derive improved information from the image.

Components of the system must be appropriate for the task at hand. This means that the interpreter must not only know intimately the remote sensing system, but also the subject of the interpretation, to include the amount of detail required, appropriate time of year to acquire the data, best spectral regions to use, and so on. Like the physical components of the system, the interpreter's knowledge and experience also interact to form a whole.

1.4.8 Role of Atmosphere

All energy reaching the remote sensing instrument must pass through a portion of the Earth's atmosphere. For satellite remote sensing in the visible and near infrared, the energy received by the sensor must pass through a considerable depth of the Earth's atmosphere. In doing so, the sun's energy is altered in intensity and wavelength by particles and gases in the Earth's atmosphere. These changes appear on the image in ways that degrade image quality or influence the accuracy of interpretations.

1.4.9 Passive and Active Remote Sensing

So far, throughout this chapter, we have made various references to the sun as a source of energy or radiation. The sun provides a very convenient source of energy for remote sensing. The sun's energy is either reflected, as it is for visible wavelengths, or absorbed and then reunited, as it is for thermal infrared wavelengths. Remote sensing systems which measure energy that is naturally available are called passive sensors. Passive sensors can only be used to detect energy when the naturally occurring energy is available. For all reflected energy, this can only take place during the time when the sun is illuminating the Earth. There is no reflected energy available from the sun at night. Energy that is naturally emitted (such as thermal infrared) can be detected day or night, as long as the amount of energy is large enough to be recorded.

Active sensors, on the other hand, provide their own energy source for illumination (Figure 1.6). The sensor emits radiation, which is directed toward the target to be investigated. The radiation reflected from that target is detected and measured by the sensor. Advantages for active sensors include the ability to obtain measurements anytime, regardless of the time of day or season. Active sensors can be used for examining wavelengths that are not sufficiently provided by the sun, such as microwaves, or to better control the way a target is illuminated. However, active systems require the generation of a fairly large amount of energy to adequately illuminate targets. Some examples of active sensors are a laser sensor and synthetic aperture radar (SAR).

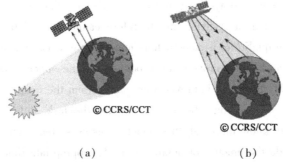

(a)   (b)

Figure 1.6 Passive RS (a), and Active RS (b)

1.5 Applications of RS

Applications of remotely sensed data to practical problems usually require links to other kinds of information, including, for example, topography, political boundaries, and pedologic, geologic, or hydrologic data. In recent years, such links have increasingly been made within the framework of GIS. Although there is no Sharp line between remote sensing and GIS, it is only partially incorrect to state that remote sensing is primarily a means of collecting data, and that GIS are primarily a means of storing and analyzing data.

At local levels of examination, remote sensing imagery records, large-scale representations of topography and drainage and the basic infrastructure of highways, buildings, and utilities. At country and regional levels of detail, remotely sensed data provide a basis for outlining broad-scale patterns of development; to coordinate the relationships between transportation, residential, industrial and recreational land uses; to site landfills, and to plan future development. State governments require information from remotely sensed data for broad-scale inventories of natural resources; to monitor environmental issues, including land reclamation and water quality; and to plan economic development.

At broader levels of examination, national governments apply the methods of remote sensing and image analysis for environmental monitoring (both domestically and internationally), for the management of federal lands, crop forecasting, disaster relief, and to support activities to further national security and international relations. Remote sensing is also used to support activities of international scope, including analysis of broad-scale environmental issues, international development, disaster relief, aid for refugees, and investigation of environmental issues of global scope.

Within subject-area disciplines, remote sensing imagery is almost always combined with other kinds of data. Geologists and geophysicists use remotely sensed images to study lithologies, structures, surface processes, and geologic hazards. Hydrologists examine images that show land cover patterns, soil moisture status, drainage systems, sediment content of lake and rivers, ocean currents, and other characteristics of water bodies. Geographers and planners examine imagery to study settlement patterns, inventory land resources, and track changes in human uses of the landscape. Foresters use remote sensing and GIS to map timber stands, estimate timber volume, monitor insect infestations, fight forest fires, and plan harvesting of timber. Agricultural scientists can examine the growth, maturing, and harvesting of crops, and monitor the progress of diseases, infestations, and droughts to forecast their impact on crop yields. Soil scientists use remotely sensed imagery to plot the boundaries of soil units and to examine relationships between soil patterns and those of land use and vegetation. In brief, remotely sensed imagery has found applications in virtually all fields that require the analysis of distributions of natural or human resources on the Earth's surface.

☞ **Key Points**

- Definitions of remote sensing
- Categories of remote sensors

- Spatial resolution, spectral resolution, radiometric resolution and temporal resolution, passive remote sensing, active remote sensing

☞ Review Questions

(1) All remotely sensed images, observe the Earth from above. The advantages to this overhead view were discussed in class, can you list some disadvantages?

(2) Remotely sensed images show the combined effects of many landscape elements, including vegetation, topography, illumination, soil, and drainage, ect. Is this diverse combination an advantages or disadvantages? Explain.

(3) Think about how spatial resolution, spectral resolution, and radiometric resolution are interrelated. Is it possible to increase one kind of resolution without influencing the others?

Chapter 2   Electromagnetic Spectrum

With the exception of objects at absolute zero, all objects emit electromagnetic radiation. Objects also reflect radiation that has been emitted by other objects. By recording emitted or reflected radiation, and applying a knowledge of its behaviour as it passes through the Earth's atmosphere and interacts with objects, remote sensing analysts develop a knowledge of the character of features such as vegetation, structures, soils, rock, or water bodies on the Earth's surface (Figure 2.1).

Interpretation of remote sensing imagery depends on a sound understanding of electromagnetic radiation and its interaction with surfaces and the atmosphere.

The discussion of electromagnetic radiation builds a foundation to permit development in subsequent lectures on the many other important topics within the field of remote sensing.

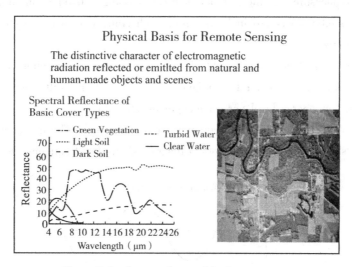

Figure 2.1   Spectra of several basic cover types

2.1   Electromagnetic Waves

Electromagnetic radiation is composed of Electric (E) and Magnetic (H) components. The electric and magnetic components are oriented at right angles to one another, and vary along an axis perpendicular to the axis of propagation (Figure 2.2).

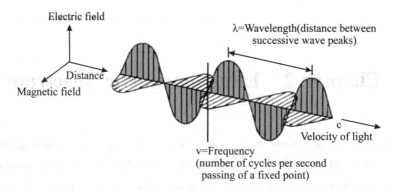

Figure 2.2 Electromagnetic waves

2.1.1 Three Properties of EV

Wavelength is the distance from one wave crest to the next. The wavelength can be measured in everyday units of length, although very short wavelengths have such small distances between wave crests that extremely short measurement units (nm, Å) are required.

Frequency is measured as the number of crests passing a fixed point in a given period of time. Frequency is often measured in hertz, units each equivalent to one cycle per second, and multiples of the hertz.

Amplitude is equivalent to the height of each peak. Amplitude is often measured as energy levels (formally known as spectral irradiance), expressed as watts per square meter per micrometer (i. e. , as energy level per wavelength in interval).

Figure 2.3 shous the amplitude, frequency, and wavelength. The center diagram represents high frequency, short wavelength; the bottom diagram shows low frequency, long wavelength.

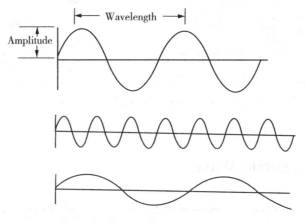

Figure 2.3 Amplitude, frequency, and wavelength

The speed of electromagnetic energy (c) is constant at 299,893 kilometers (km) per second. Frequency (v) and wavelength (λ)

$$c = \lambda v$$

The frequency is inversely proportional to wavelength, which means that the longer the wavelength, the lower the frequency or the shorter the wavelength, the higher the frequency.

The characteristics of electromagnetic energy can be specified using either frequency or wavelength. Varied disciplines, and varied applications, follow different conventions for describing electromagnetic radiation, using either wavelength, or frequency.

Although there is no authoritative standard, a common practice in the field of remote sensing is to define regions of the spectrum.

2.1.2 Radiation Laws

1. Stefan-Boltzmann Law

All of the Earth objects—water, soil, rocks, and vegetation—above absolute zero (0 K) and the sun, emit electromagnetic energy. The sun is the initial source of electromagnetic energy that can be recorded by many remote sensing systems; radar and sonar systems are exceptions, as they generate their own source of electromagnetic energy. The amount of energy and the wavelengths at which it is emitted depend upon the temperature of the object, as the temperature of an object increases, the total amount of energy emitted also increases, and the wavelength to maximum (peak) emission becomes shorter. These relationships can be expressed formally using the concept of the black body. A black body is defined as an object that totally absorbs and emits radiation at all wavelengths. For example, the sun is a 6,000 K black body. The total emitted radiation from a black body (M_λ) is measured in watts per m^2 (W/m^2) and is proportional to the fourth power of its absolute temperature (T), measured in degrees Kelvin:

$$M_\lambda = \sigma T^4$$

Where: σ is the Stefan-Boltzmann constant of 5.6697×10^{-8} W · m^{-2} · K^{-4}.

The law states that the amount of energy emitted by an Earth's object (or the Sun) is a function of its temperature. The higher the temperature, the greater the amount of radiant energy emitted by the object.

2. Wien's Displacement Law

The dominant wavelength of the radiation can be calculated as:

$$\lambda_{max} = k/T$$

Where: k is a constant of 2,898 and T is the absolute temperature, degrees K.

For example, we can calculate the dominant wavelength of the radiation for the Sun at 6,000 K and the Earth at 300 K as:

$\lambda_{max} = 2898\ \mu m \cdot K / 6000\ K$

$\lambda_{max} = 0.483\ \mu m$

&

$\lambda_{max} = 2898 \ \mu m \cdot K/300 \ K$

$\lambda_{max} = 9.66 \ \mu m$

Figure 2.4 shows the black body radiation curves for several objects, including the Sun and the Earth, which approximate 6,000 K and 300 K black bodies, respectively. The area under each curve may be summed to compute the total radiant energy (M_t) exiting each object. Thus, the Sun produces more radiant exitance than the Earth because its temperature is greater. As the temperature of an object increases, its dominant wavelength (λ_{max}) shifts toward the shorter wavelengths of the spectrum.

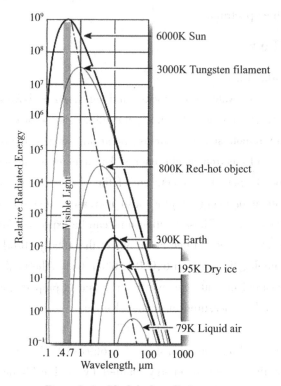

Figure 2.4 Black body radiation curves

3. Kirchhoff's Law

The black body is a hypothetical entity because in nature all objects reflect at least a small proportion of the radiation that strikes them, and thus do not act as perfect reradiators of absorbed energy. Although truly perfect black bodies cannot exist, their behavior can be approximated using laboratory instruments. Such instruments have formed the basis for the scientific research that has defined the relationships between the temperatures of objects and the radiation they emit.

Kirchhoff's law states that the ratio of emitted radiation of emitted radiation of absorbed radiation flux is the same for all black bodies at that same temperature. This law forms the basis for the definition of emissivity (ε), the ratio between the emittance of a given object (M) and that of black body at the same temperature (M_b):

$$\varepsilon = M/M_b$$

The emissivity of a true black body is 1, and that of a perfect reflector (a white body) would be 0. Black bodies and white bodies are hypothetical concepts, approximated in the laboratory under contrived conditions. In nature, the emissivity of all objects that falls between these extremes (gray bodies). For these objects, emissivity is a useful measure of their effectiveness as radiators of electromagnetic energy. Those objects that tend to absorb high proportions of incident radiation and then to reradiate this energy will have high emissivities. Those that are less effective as absorbers and radiators of energy have low emissivities (i. e., they return much more of the energy that reaches them).

2.2 The Spectral Region Used in RS

The electromagnetic (EM) spectrum is the continuous range of electromagnetic radiation, extending from gamma rays (the highest frequency & shortest wavelength) to radio waves (the lowest frequency & longest wavelength) and including visible light.

The EM spectrum can be divided into seven different regions—gamma rays, X-rays, ultraviolet, visible light, infrared, microwaves and radio waves (Figure 2.5).

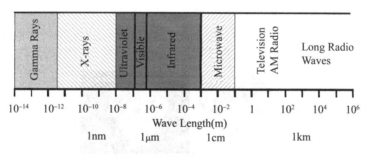

Figure 2.5 EM spectrum

2.2.1 Ultraviolet (UV)

The UV region ranges from 0.03 μm to 0.4 μm micrometers. These are the shortest wavelengths practical for remote sensing.

The UV can be divided into the far UV (0.01~0.2 μm), middle UV (0.2~0.3 μm) and near UV (0.3~0.4 μm).

The sun is the natural source of UV radiation. However, wavelengths that shorter than 0.3 μm are unable to pass through the atmosphere and reach the Earth's surface due to atmospheric absorption by the ozone layer. Only the 0.3~0.4 μm wavelength region is useful for terrestrial remote sensing. The near UV radiation is known for its ability to induce fluorescence, emission of visible radiation, in some materials; it has significance for a specialized form of remote sensing.

2.2.2 Visible (V)

The most common region of the electromagnetic spectrum used in remote sensing is the visible band, which spans from 0.4~0.7 μm. These limits correspond to the sensitivity of the human eye. Blue (0.4~0.5 μm), green (0.5~0.6 μm), and red (0.6~0.7 μm) represent the additive primary colors-colors that cannot be made from any other colors.

Although sunlight seems to be uniform and homogeneous in color, sunlight is actually composed of various wavelengths of radiation, primarily in the UV, visible and near IR portions of the electromagnetic spectrum.

Most of the colors that we see are a result of the preferential reflection and absorption of wavelengths that make up white light.

For example, chlorophyll in healthy vegetation selectively absorbs the blue and red wavelengths of white light to use in photosynthesis and reflects more of the green wavelengths. Thus, vegetation appears as green to our eyes.

Snow is seen as white, since all wavelengths of visible light are scattered.

The three primary colors reflected from an object (Figure 2.6) known as *additive primaries* are the blue, green and red wavelengths. Primary colors cannot be formed by the combination of any other colors.

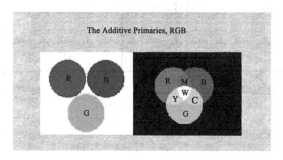

Figure 2.6 The additive primary colors

Intermediate colors are formed when a combination of primary colors is reflected from an object. For example, magenta is a combination of reflected red and blue, cyan a combination of reflected blue and green, and yellow a combination of reflected red and green.

2.2.3 Infrared (IR)

The IR band includes wavelengths between the red light (0.7 μm) of the visible band and microwaves at 1,000 μm. Infrared literally means "below the red" because it is adjacent to red light.

The reflected near IR region (0.7~1.3 μm) is used in black and white IR and color IR sensitive film.

The middle IR includes energy at wavelengths ranging from 1.3 μm to 3 μm. Middle IR energy is detected using electro-optical sensors.

The thermal (far) IR band extends from 3 μm to 1,000 μm. However, due to atmospheric attenuation, the wavelength regions of 3 ~ 5 μm and 8 ~ 14 μm are typically used for remote sensing studies.

The thermal IR region is directly related to the sensation of heat. Heat energy, which is continuously emitted by the atmosphere and all features on the Earth's surface, dominates the thermal IR band. Optical-mechanical scanners and special vidicon systems are typically used to record energy in this part of the electromagnetic spectrum.

2.2.4 Microwave (MV)

The microwave region is from 1 mm to 1 m. Microwaves can pass through clouds, precipitation, tree canopies and dry surficial deposits.

There are two types of sensors that operate in the microwave region. Passive microwave systems detect natural microwave radiation that is emitted from the Earth's surface. Radar (radio detection and ranging) systems-active remote sensing-propagate artificial microwave radiation to the Earth's surface and record the reflected component.

Figure 2.7 is the bands used in remote sensing.

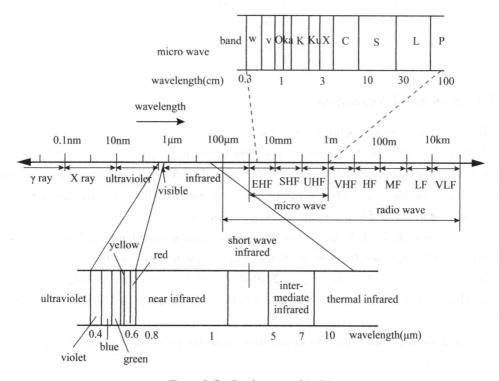

Figure 2.7 Bands are used in RS

2.3 Typical EMR Spectra

Figure 2.8 shows the spectra the of several materials

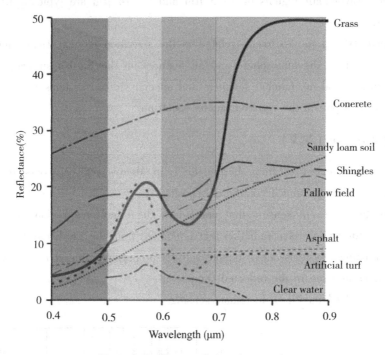

Figure 2.8 Spectra of several materials

2.3.1 Spectra of Vegetation

Figure 2.9 shows the spectra of vegetation.

Visible (0.4~0.7 μm): low reflectance, low transmittance, high absorption mainly due to chlorophyll centered in the blue (0.45 μm) and the red (0.67 μm) wavelength zone, although other leaf pigments like xanthophylls, carotenoids, and anthocyanins also affect absorption, and small peak centered at 0.55 μm in the yellow-green region.

Near-infrared (0.7~1.3 μm): low absorption, high transmittance as leaf pigments and cellulose of cell walls are transparent. Near-infrared plateau between 0.7 and 1.3 μm and near-infrared edge around 0.7 μm. Figure 2.10 shows the location of the red edge.

Middle-infrared (1.3~2.5 μm): strong water absorption bands at 1.4 μm, 1.9 μm, and 2.7 μm.

Figure 2.11 shows the effect of moisture content on the reflectance of corn leaves.

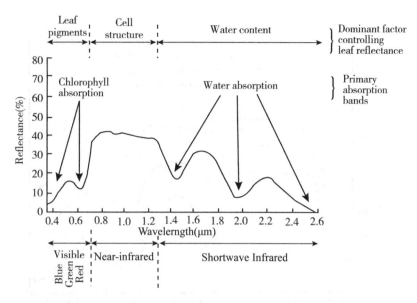

Figure 2.9 Spectra of vegetation

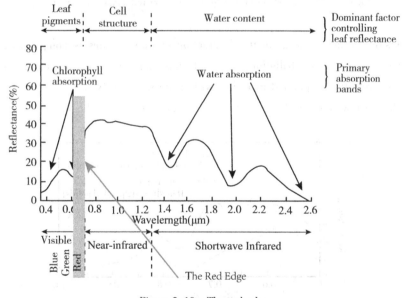

Figure 2.10 The red edge

2.3.2 Spectra of Water

Interaction of EMR with water is a function of the nature of water and its conditions.

Delineation of water bodies can be done more easily in the NIR wavelengths.

In a natural setting, water bodies absorb nearly all incident energy in both NIR and MIR wavelengths.

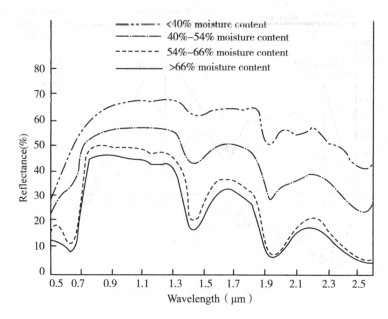

Figure 2.11 The effect of moisture content on reflectance of corn leaves

In the visible portion, energy-matter interaction of water bodies is complex. (For example, the reflectance may involve contributions from reflectance, from surface of water, from the bottom of materials, or from the suspended materials within the water body).

Figure 2.12 shows the spectra of turbid river water and clear lake water.

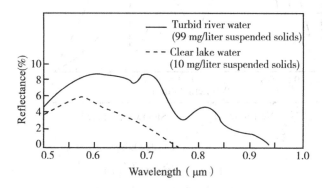

Figure 2.12 Spectra of water

2.3.3 Spectra of Snow

It is difficult to differentiate between snow and cloud due to similarity in spectral response in wavelength 0.5 to 1.1 μm.

Mapping the extent and conditions of snow covering can be best done in the middle IR bands. A typical snow spectra is shown in Figure 2.13.

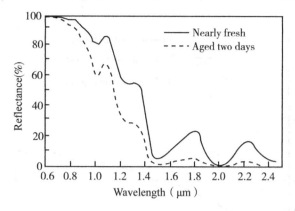

Figure 2.13 Spectra of snow

2.3.4 Spectra of Soil

The following main factors affect the interaction of EMR with soil: moisture content, organic matter, mineralogy or chemical composition, particle size and texture. Figure 2.14 shows the spectra of several types of soil.

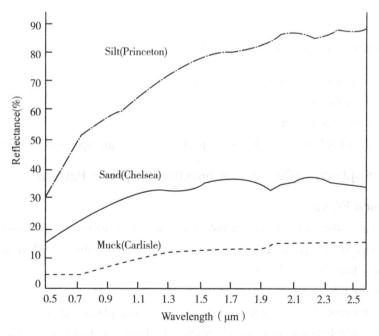

Figure 2.14 Spectra of several types of soil

Figure 2. 15 shows the effect of the moisture content on the soil spectra.

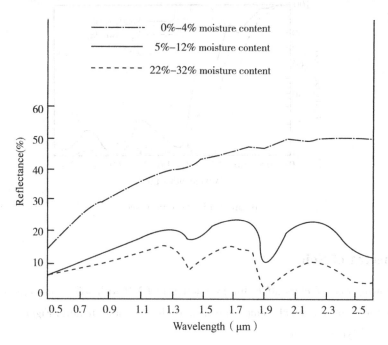

Figure 2. 15 The effect of the moisture content on the soil spectra

2.3.5 How do you get the spectra?

- Measure in the field with field spectrometers
- Measure in the lab
- Collect from image data
- Look at spectral libraries: http: //speclab. cr. usgs. gov/spectral-lib. html

2.3.6 Atmospheric Influences on Spectral Response Patterns

1. Temporal Effects

Factors change the spectral characteristics of a feature over time. For example, response of vegetation throughout the growing season. These changes often influence when we might collect sensor data for a particular application.

2. Spatial Effect

Factors that cause the same type of features (e. g. corn plants) at a given point in time to have different characteristics at different geographic locations (different soils, climate, and cultivation practices).

2.4 Interactions with the Atmosphere

Figure 2.16 shows the schematic solar radiation interactions with the atmosphere. All radiation used for remote sensing must pass through the Earth's atmosphere. If the sensor is carried by a low-flying aircraft, the effects of the atmosphere upon image quality may be negligible. In contrast, energy that reaches sensors carried by earth satellites must pass through the entire depth of the Earth's atmosphere. Under these conditions, atmospheric effects may have substantial impact upon the quality of images and data that the sensors generate. Therefore, the practice of remote sensing requires knowledge of interactions of electromagnetic energy in the atmosphere. Figure 2.17 shows several interactions with the atmosphere before reaching the sensor.

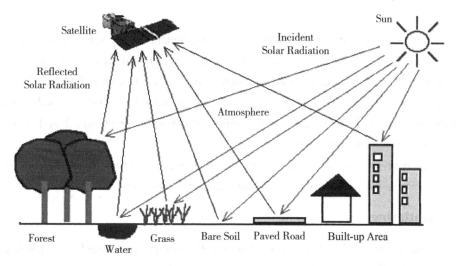

Figure 2.16 Interactions between solar radiation and atmosphere

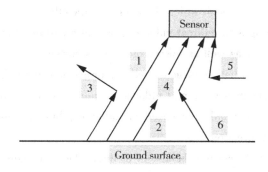

Figure 2.17 Interactions of solar energy with atmosphere

Ray-1: Photons which leave the Earth's surface and reach the sensor without change. This constitutes useful signal for remote sensing.

Ray-2: Photons which leave the Earth's surface heading in the direction of the sensor, but which are absorbed by interaction with the atmosphere en route.

Ray-3: Photons diverted out of the sensor's field of view (FOV) by scattering as a result of atmosphere interaction.

Ray-4: Photons of EM energy which are emitted by the atmosphere itself.

Ray-5: Photons of energy from the illuminating source (sun or active radar source) which are scattered into the FOV of the sensor without touching the Earth's surface (land/sea target).

Ray-6: Photons which have left the ground and carry information from an area other than the ground FOV of the sensor, and which are deflected by the atmospheric scattering into the FOV of the sensor.

Measure the transmittance of the atmosphere (transmittance = component 1/components (1+2+3)).

If the energy making up ray-1 (the remote sensing signal) is very small in comparison to rays 4, 5, and 6 (the atmospheric contribution) then the possibilities of remote sensing are reduced. However, by following a proper atmospheric correction algorithm, it should be possible to enhance the ground signal. The atmospheric correction algorithm must not only remove the contribution of rays 4, 5, and 6, but also estimate the magnitude of rays 2 and 3 in order to recover the true ground-leaving signal.

As solar energy passes through the Earth's atmosphere, it is subject to modification by several physical processes, including scattering, absorption, and refraction, etc.

2.4.1 Scattering

Scattering is the redirection of electromagnetic energy by particles suspended in the atmosphere or by large molecules of atmospheric gases. The amount of scattering that occurs depends upon the sizes of these particles, their abundance, the wavelength of the radiation, and the depth of the atmosphere through which the energy is travelling.

The effect of scattering is to redirect radiation, so that a portion of the incoming solar beam is directed back toward space, as well as toward the Earth's surface.

Once electromagnetic radiation is generated, it propagates through the Earth's atmosphere almost at the speed of light in a vacuum. Unlike a vacuum in which nothing happens, however, the atmosphere may affect not only the speed of radiation, but also its wavelength, intensity, spectral distribution, and direction.

Figure 2.18 shows the scattering behaviors of three classes of atmospheric particles. Figure 2.18(a) shows that the atmospheric dust and smoke form rather large irregular particles that create a strong forward-scattering peak, with a smaller degree of backscattering. Figure 2.18(b) shows that the atmospheric molecules are more nearly symmetric in shape, creating a pattern characterized by preferential forward and backscattering, but without the pronounced peaks observed in the first example. Figure 2.18 (c) shows that the large water droplets create a pronounced forward-scattering peak, with smaller backscattering peaks.

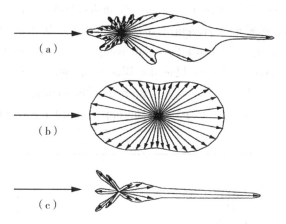

Figure 2.18 Three types scattering behaviors

There are essentially three types of scattering (Figure 2.19): Rayleigh, Mie, and Nonselective scattering. The occurrence of scattering type mainly restricted to (a) the wavelength of the incident radiant energy, and (b) the size of the gas molecule, dust particle, and/or water vapour droplet encountered.

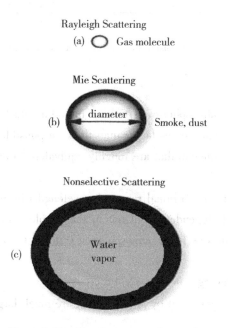

Figure 2.19 The three types of scattering

1. Rayleigh Scattering

A clean atmosphere, consisting only of atmospheric gases, causes scattering of light in a manner such that the amount of scattering increases greatly as the wavelength becomes shorter. Rayleigh scattering is the most common scattering produced by atmospheric gases. It occurs when the diameter of the matter (usually air molecules) is many times smaller than the wavelength of the incident electromagnetic radiation ($a \ll \lambda$). It depends on the wavelength a proportional to $1/\lambda^4$. Figure 2.20 shows the relationship between intensity of Rayleigh scattering and wavelength. It is responsible for the blue sky. Blue light (400 nm) is scattered 16 times more than near-infrared light (800 nm). Rayleigh scattering is the dominant scattering process high in the atmosphere, up to altitudes of 9 to 10 km, the upper limit for atmospheric scattering.

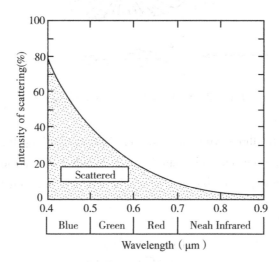

Figure 2.20 The relationship between intensity of scattering and wavelength

2. Mie Scattering

It is caused by large atmospheric particles, including dust, pollen, smoke, and water droplets. Such particles are many times larger than those responsible for Rayleigh scattering.

Those particles have diameters that are roughly equivalent to the wavelength of the scattered radiation.

Mie scattering can influence a broad range of wavelengths in and near the visible spectrum; Mie scattering is wavelength-dependent, but not in the simple manner of Rayleigh scattering; it tends to be the greatest in the lower atmosphere (0 to 5 km), where larger particles are abundant.

3. Nonselective Scattering

Nonselective scattering is caused by particles that are much larger than the wavelength of the scattered radiation.

For radiation in and near the visible spectrum, such particles might be larger water droplets or large particles of airborne dust.

"Nonselective" means the scattering is not wavelength-dependent, so we observe it as a whitish or grayish haze—all visible wavelengths are scattered equally.

2.4.2 Refraction

Refraction is the bending of light rays in the contact area between two media that transmit light. Refraction also occurs in the atmosphere as light passes through atmospheric layers of varied clarity, humidity, and temperature. These variables influence the density of atmospheric layer, which in turn causes a bending of light rays as they pass from one layer to another.

An example is the shimmering appearances on hot summer days of objects viewed in the distance as light passes through hot air near the surface of heated highways, runways, and parking lots.

2.4.3 Absorption

Absorption is the process by which radiant energy is absorbed and converted into other forms of energy.

Absorption occurs when the energy of the same frequency as the resonant frequency of an atom or molecule is absorbed producing an excited state. Instead of re-radiating a photon of the same wavelength, the energy is transformed into heat motion and is reradiated at a longer wavelength, absorption occurs.

An absorption band is a range of wavelengths (or frequencies) in the electromagnetic spectrum within which radiant energy is absorbed by substances such as water (H_2O), carbon dioxide (CO_2), oxygen (O_2), ozone (O_3), and nitrous oxide (N_2O).

The cumulative effect of the absorption of the various constituents can cause the atmosphere to close down in certain regions of the spectrum (Figure 2.21). This is bad for remote sensing because no energy is available to be sensed.

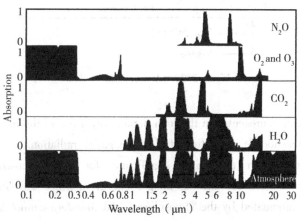

Figure 2.21 Cumulative effect of the absorption

The Sun's incident electromagnetic energy can be absorbed by all kinds of atmospheric gas molecules in the range of 0.1 to 30 μm. Ozone plays an important role in the Earth's energy balance. Absorption of the high-energy, short-wavelength, portions of the ultraviolet spectrum prevents transmission of this radiation to the lower atmosphere. Ozone absorbs strongly in the UV (short wavelengths) and protects us from skin cancer!

Carbon dioxide is important in remote sensing because it is effective in absorbing radiation in the mid-and far-infrared regions of the spectrum. Its strongest absorption occurs in the region from about 13 to 17.5 μm, in the mid infrared. It is the reason for the greenhouse effect.

The abundance of water vapor varies greatly from time to time and from place to place. Water vapor is several times more effective in absorbing radiation than are all other atmospheric gases combined. Two of the most important regions of absorption are in several bands between 5.5 and 7.0 μm, and above 27.0 μm.

1. Atmospheric Windows

Those wavelengths that are relatively easily transmitted through the atmosphere are referred to as atmospheric windows (Figure 2.22).

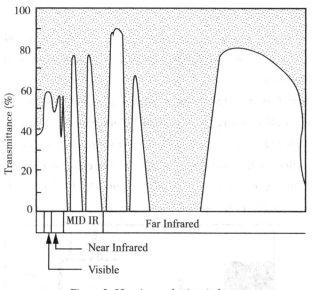

Figure 2.22　Atmospheric windows

This is a schematic representation that can depict only a few of the most important windows. The shaded area represents the absorption of electromagnetic radiation.

Atmospheric windows are of obvious significance for remote sensing—they define those wavelengths that can be used for forming images energy at other wavelengths, not within the windows, is severely attenuated by the atmosphere, and therefore cannot be effective for remote sensing.

In the far-infrared region, the two most important windows extend from 3.5 to 4.1 μm, and from 10.5 to 12.5 μm. The latter is especially important because it corresponds approximately to wavelengths of peak emission from the Earth's surface. Table 2.1 shows the major atmospheric windows.

Table 2.1   **Major atmospheric windows used in RS**

Ultraviolet and visible	0.30~0.75μm
	0.77~0.91μm
Near infrared	1.55~1.75μm
	2.05~2.4μm
Thermal infrared	8.0~9.2μm
	10.2~12.4μm
Microwave	7.5~11.5μm
	20.0+mm

2. Clouds

Most EMR wavelengths cannot penetrate clouds. It is a big problem in remotely sensed imagery—tropics especially. Generally speaking, temporal compositing is often used to get rid of the clouds. Cloud shadow is also a problem.

2.5 Interactions with Surfaces

As electromagnetic energy reaches the Earth's surface, it must be reflected, absorbed, or transmitted. The proportions accounted for by each process depend upon the nature of the surface, the wavelength of the energy, and the angle of illumination.

2.5.1 Reflectance

1. Definition

The amount of reflected radiation divided by that amount of incoming radiation in a particular wavelength, which can be expressed as:

Reflectance (%) = reflected/irradiance · 100%

2. Types of Reflection

If the surface is smooth relative to wavelength, specular reflection occurs. Specular reflection redirects all, or almost all, of the incident radiation in a single direction. For such surfaces, the angle of incidence is equal to the angle of reflection. For visible radiation, specular reflection can occur with surfaces such as a mirror, smooth metal, or a calm water body.

If a surface is rough relative to wavelength, it acts as a diffuse, or isotropic, reflector. Energy is scattered more or less equally in all directions. For visible radiation, many natural surfaces might behave as diffuse reflectors, for example, uniform grassy surfaces. A perfectly diffuse reflector (known as a Lambertian surface) would have equal brightnesses when observed from any angle.

Most Earth surfaces are neither perfectly specular nor perfectly diffuse reflectors.

The category that describes any given surface is dictated by the surface's roughness in comparison to the wavelength of the energy being sensed.

Diffuse reflection contains spectral information on the colour of the reflecting surface, whereas specular reflection generally not.

In Remote Sensing, we are most often interested in measuring the diffuse reflection properties of terrain features.

Reflectance varies with wavelength, geometry and it is diagonstic of different materials.

2.5.2 Transmittance

Transmission of radiation occurs when radiation passes through a substance without significant attenuation. From a given thickness, or depth, of a substance, the ability of a medium to transmit energy is measured as the transmittance (t):

$$t = \frac{\text{Transmitted Radiation}}{\text{Incident Radiation}}$$

In the field of remote sensing, the transmittance of films and filters is more important. With respect to naturally occurring materials we often think only of water bodies as capable of transmitting significant amounts of radiation. However, the transmittance of many materials varies greatly with wavelengths, so our direct observations in the visible spectrum do not transfer to other parts of the spectrum. For example, plant leaves are generally opaque to visible radiation but transmit significant amounts of radiation in the infrared.

☞ Summary

- Electromagnetic spectrum
- Radiation laws
- EMR spectra
- Interactions with atmosphere and surface

☞ Key Points

- Three properties of EV
- Stefan-Boltzmann law
- Wien's displacement law
- Kirchhoff's laws

- Spectra of vegetation and water
- Scattering
- Refraction
- Reflectance
- Atmospheric windows

☞ **Review Questions**

(1) Rayleigh scattering is the cause both for the blue color of the sky and for the brilliant red and orange colors often seen at sunset. Why?

(2) Some streetlights are deliberately manufactured to provide illumination with a reddish color. From material presented in this chapter, can you suggest why?

Chapter 3 Platforms and Sensors

3.1 Introduction

The WMO Global Observing System (GOS) enables the observation and collection of weather, water and climate information from around the globe (Figure 3.1). Through this system, data is collected from 14 satellites, hundreds of ocean buoys, aircraft, ships and nearly 10,000 land-based stations. National Meteorological and Hydrological Services (NMHSs) make and collect observations in their countries. More than 50,000 weather reports and several thousand charts and digital products are disseminated daily through the WMO Global Telecommunication System (GTS), which interconnects the National Meteorological Centers (NMSs) around the globe (Figure 3.1).

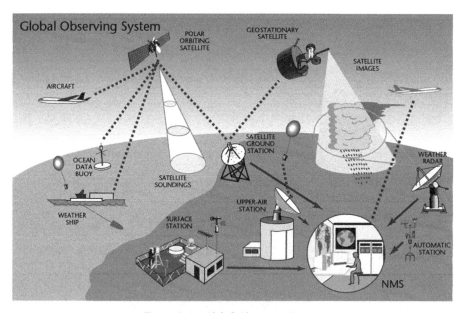

Figure 3.1 Global Observing System

3.2 Platforms

- Ground (<100 m)

- Aircraft (100 m ~ 100 km)

Low, medium, and high altitude.

Higher level of spatial detail.

- Satellite (>240 km)

Polar-orbiting and sun-synchronous: 800 ~ 900 km altitude, 90 ~ 100 minutes/orbit.

Geo-synchronous: stationary relative to Earth, 35,900 km altitude, 24 h/orbit.

3.2.1 Orbits and Swaths

The path followed by a satellite is referred to as its orbit. Satellites which view the same portion of the Earth's surface at all times have geostationary orbits. Weather and communication satellites commonly have these types of orbits. Many satellites are designed to follow a north-south orbit which, in conjunction with the Earth's rotation (west-east), allows them to cover most of the Earth's surface over a period of time. These are near-polar orbits. Many of these satellite orbits are also Sun-synchronous, such that they cover each area of the world at a constant local time of day. Near-polar orbits also mean that the satellite travels northwards on one side of the Earth and then toward the southern pole on the second half of its orbit. These are called ascending and descending passes, respectively.

As a satellite revolves around the Earth, the sensor sees a certain portion of the Earth's surface. The area imaged is referred to as the Swath. The surface directly below the satellite is called the nadir point. Steerable sensors on satellites can view an area (off nadir) before and after the orbits passes over a target.

3.2.2 Sun-synchronous Satellites

Satellites orbit that remains constant in relation to the Sun.

1. Polar Orbit

A polar orbit is a particular type of Low Earth Orbit. The only difference is that a satellite in polar orbit travels a north-south direction, rather than the more common east-west direction.

2. Why use the Polar Orbit

Polar orbits are useful for viewing the planet's surface. As a satellite orbits in a north-south direction, Earth spins beneath it in an east-west direction. As a result, a satellite in polar orbit can eventually scan the entire surface.

3. Inclinations

We say that a polar orbit has an inclination, or angle, of 90 degrees. It is perpendicular to an imaginary line that slices through Earth at the equator.

4. Low Earth Orbit

When a satellite circles close to Earth, we say it's in Low Earth Orbit (LEO). Satellites in LEO are just 200 ~ 500 miles (320 ~ 800 kilometers) high. Because the orbit is so close to the Earth, they must travel very fast so gravity won't pull them back into the atmosphere. Satellites in

LEO speed along at 17,000 miles per hour (27,359 kilometers per hour). They can circle the Earth in about 90 minutes.

5. What a View

A Low Earth Orbit is useful because its nearness to the Earth gives it spectacular views. Satellites that observe our planet, like Remote Sensing and Weather satellites, often travel in LEOs because from this height they can capture very detailed images of the Earth's surface.

6. Space Junk

The LEO environment is getting very crowded. The United States Space Command (USSC) keeps track of the number of satellites in orbit. According to the USSC, there are more than 8,000 objects larger than a softball.

Polar orbits allow for lower altitudes and more image resolution. These satellites take multiple passes of the Earth before returning to the same location.

3.2.3 Geostationary satellites

Geostationary satellites are able to view the Earth from above. As they move in synchronicity with the Earth's rotation, they can provide regular coverage for a region and help in forecasting.

Geostationary orbits are circular orbits that are orientated in the plane of the Earth's equator. By placing the satellite at an altitude where its orbital period exactly matches the rotation of the Earth (35,800 km), the satellite appears to "hover" over one spot on the Earth's equator. While geostationary satellite is ideal for making repeated observations of a fixed geographical area centered on the equator, they are far enough away from the Earth to make it difficult to obtain high quality, quantitative observations. The current generations of geostationary meteorological satellites are surely technological marvels. These satellites, however, do not see the poles at all, and to get global coverage of just the equatorial regions, you need a network of 5~6 satellites.

3.3 Remote Sensing Data Sources

3.3.1 Coarse Resolution Remotely Sensed Data

1. AVHRR (Advanced Very High Resolution Radiometer)

Originally intended for meteorological work, but have proved useful for many applications.

2. Spectral Resolution
- 0.58~0.68 μm (Red)
- 0.725~1.10 μm (Near IR)
- 3.55~3.93 μm (Thermal IR)
- 10.3~11.3 μm (Thermal IR)
- 11.5~12.5 μm (Thermal IR)

Advantage: Frequent global coverage (high temporal resolution), less data processing/

unit².

 Disadvantage: Coarse spatial resolution.

3. Applications of AVHRR

Vegetation monitoring and desertification:

- Vegetation Index = Band 2-Band 1
- Normalized Difference Vegetation Index (NDVI) = (Band2-Band1)/(Band2+Band1)

Global change studies:

Climate change/Monitoring

3.3.2 Medium Resolution Satellites

1. EOS

Earth Observing System (EOS): part of the Earth Science the Enterprise—a NASA initiative to assess the impacts of natural events and human-induced activities in the Earth's environment. Includes both space and ground-based measurement systems.

Five sensors on the EOS Terra satellite:

ASTER—Visible/Near IR sensors, along track stereo, produce DEM's.

CERES—Radiation balance measurements (global warming).

MISR—Multiangle views of the Earth, atmospheric data.

MODIS—"New & improved AVHRR"—36 channels (bands), 250 m, 500 m, or 1,000 m spatial resolution.

MOPITT—Measurement of Pollution in the Troposphere.

2. MODIS

- Moderate Resolution Imaging Spectroradiometer
- Instrument aboard Terra Satellite
- Data in 36 spectral bands 0.4 to 14.4 μm
- Resolution: 2 bands at 250 m, 5 bands at 500 m, and 29 bands at 1,000 m
- Orbit 705 km
- Swath width 2,330 km
- Global coverage 1~2 days
- Website: http://modis.gsfc.nasa.gov

3. ASTER

- Advanced Spaceborne Thermal Emission and Reflection Radiometer
- Instrument aboard Terra Satellite
- Has 3 subsystems: 3 visible and near-infrared (VNIR) bands with 15 m resolution, 6 shortwave infrared (SWIR) bands with 30 m resolution, 5 thermal infrared (TIR) bands with 90 m resolution

Figure 3.2 shows the principle relationships among scientific objectives, measurements, and instruments aboard the Terra satellite.

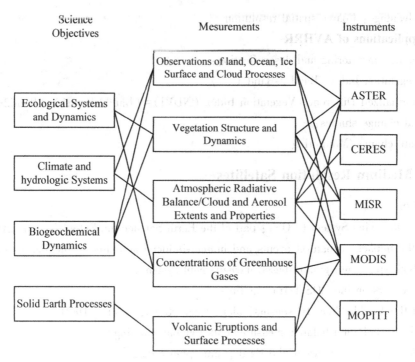

Figure 3.2 shows the principle relationships among scientific objectives, measurements, and instruments aboard the Terra satellite

4. Landsat satellites

- Seven Landsat satellites in series (so far); 3 still in operation (Landsat 4, Landsat 7, Landsat 8). Landsat 6 suffered a launch failure and was never operational
- Agency: National Aeronautics and Space Administration (NASA)
- Orbits: Polar orbit at 705 km, sun-synchronous
- Swath width: 185 km
- Repeat coverage period: 16 days

Landsats 1, 2 and 3

Earth Resources Technology Satellite (ERTS-1), later renamed Landsat 1, launched in 1972 as an experimental system to test the feasibility of collecting earth resources data from satellites.

(1) Data made publicly available world-wide: Open Skies Policy.

(2) Carried a multispectral scanner (MSS), imaged a 185 km swath in 4 wavebands, 2 in visible, 2 in near-infrared, spatial resolution of 80 m, sun-synchronous orbit, repeat cycle of 18 days.

Landsats 4, and 5

(1) Landsat 4 deactivated shortly after launch, but remains in orbit.

(2) Landsat 5 carries a multispectral scanner (MSS) and a Thematic Mapper (TM), imaged a 185 km swath.

(3) Seven wavebands from the visible blue to the thermal infrared, spatial resolution of 30m except the thermal band (120 m), sun-synchronous orbit, repeat cycle of 16 days.

(4) Each scene contains about 36 million pixels and 250 million data values.

Table 3.1 shows the characteristics of TM sensor.

Table 3.1　　　　　　　　　　**TM Sensor characteristics**

Band	Wavelength (μm)	Spectral Location	Resolution(m)
1	0.45~0.52	Blue	30
2	0.53~0.60	Green	30
3	0.63~0.69	Red	30
4	0.76~0.90	Near IR	30
5	1.55~1.75	Mid IR	30
6	10.4~12.5	Thermal IR	120
7	2.08~2.35	Mid IR	30

Band 1 is mainly used for coastal water mapping, soil/vegetation discrimination, forest type mapping, and cultural feature identification.

Band 2 is mainly used for measuring green reflectance peak of vegetation for vegetation discrimination, vigor assessment, and cultural feature identification.

Band 3 is mainly used for sensing a chlorophyll absorption region, aiding in plant species differentiation, cultural feature identification.

Band 4 is mainly used for determining vegetation types, vigor and biomass content, delineate water bodies, soil moisture discrimination.

Band 5 is mainly used for estimating vegetation moisture content & soil moisture, differentiating snow from clouds.

Band 6 is useful for vegetation stress analysis, soil moisture discrimination, thermal mapping applications.

Band 7 is mainly used for discriminating mineral from rock types, and it is sensitive to vegetation moisture content.

How the various types of resolution of Landsat imagery have changed over the years? Please see Table 3.2.

Landsat 7

(1) Landsat 7 was launched in April 1999.

(2) Landsat 7 carries a enhanced thematic mapper plus (ETM+) imaged a 185 km swath.

(3) A 15-m panchromatic channel, 7 wavebands from the visible blue to the thermal infrared, spatial resolution of 30 m except the thermal band (60 m).

Table 3.2　　　　　　　　**Sensors used on Landsat 1~7 missions**

Sensor	Mission	Sensitivity(μm)	Resolution(m)
RBV	1, 2	0.475~0.575	80
		0.580~0.680	80
		0.690~0.830	80
	3	0.505~0.750	30
MSS	1~5	0.5~0.6	79/82[a]
		0.6~0.7	79/82[a]
		0.7~0.8	79/82[a]
		0.8~1.1	79/82[a]
	3	10.4~12.6[b]	240
TM	4, 5	0.45~0.52	30
		0.52~0.60	30
		0.63~0.69	30
		0.76~0.90	30
		1.55~1.75	30
		10.4~12.5	120
		2.08~2.35	30
ETM[c]	6	Above TM bands plus 0.50~0.90	30(120m thermal band) 15
ETM+	7	Above TM Bands plus 0.50~0.90	30(60m thermal band) 15

Note: [a]79m for Landsats 1, 2, 3 and 82 m for Landsats 4, 5.

[b]Failed shortly after launch (band 8 of Landsat 3).

[c]Landsat 6 launch failure.

(4) Sun-synchronous orbit, repeat cycle of 16 days.

(5) The Scan Line Corrector (SLC) in the ETM+ instrument failed on May 31, 2003. Approximately 22% of the data on a Landsat 7 scene are missing without a functional SLC.

Landsat 8

(1) Landsat 8 was launched in February 2013.

(2) Landsat 8 carries two instruments: the Operational Land Imager (OLI) sensor and the

Thermal Infrared Sensor (TIRS). OLI includes refined heritage bands, along with three new bands: a deep blue band for coastal/aerosol studies, a short wave infrared band for cirrus detection, and a quality assessment band. TIRS provides two thermal bands which are useful in providing more accurate surface temperatures.

(3) OLI multispectral bands 1 ~ 7, 9: 30 meters; OLI panchromatic band 8: 15 meters; THIR bands 10-11: 100 meters.

(4) Approximate scene size is 170 km north-south by 183 km east-west.

(5) Landsat 8 images have a large file size of approximately 1 GB compressed.

5. SPOT Satellites
SPOTS 1, 2, and 3

SPOT—began operations in 1986, with the launch of SPOT 1. SPOT was conceived and designed by the french Centre National d'Etudes Spatiales (CNES) in Paris, with the cooperation of other European organizations. SPOT 1, launched on February 22, 1986, was followed by SPOT 2(January 22, 1990), SPOT 3 (September 26, 1993), SPOT 4 (March 24, 1998) and SPOT 5(May 3, 2002).

The SPOT consists of two identical sensing instruments, a telemetry transmitter, and magnetic tape recorders. The two sensors are known as HRV (high resolution visible) instruments. The HRV can be operated in either of two modes. In Panchromatic (PN) mode, the sensor is sensitive across a broad spectral band from 0.51 to 0.73 mm. It images a 60-km swath with 6,000 pixels per line, for a spatial resolution of 10 m. In this mode the HRV instrument provides fine spatial detail, but records a rather broad spectral region. In amother mode, the multispectral (XS) configuration, the HRV instrument senses three spectral regions:

- Band 1: 0.50~0.59 μm (green)
- Band 2: 0.61~0.68 μm (red; chlorophyll absorption)
- Band 3: 0.79~0.89 μm (near infrared; atmospheric penetration)

In this mode the sensor images a strip 60 km in width using 3,000 samples for each line, at a spatial resolution of about 20 m. Thus, in the multispectral mode, the sensor records fine spectral resolution but coarse spatial resolution. In some instances, it is possible to "sharpen" the lower spatial detail of multi-spectral images by superimposing them on the fine spatial detail of high-resolution panchromatic imagery of the same area.

With respect to sensor geometry, each of the HRV instruments can be positioned in either nadir viewing or off-nadir viewing. For nadir viewing, both sensors are oriented in a manner that provides coverage of adjacent ground segments. Because the two 60-km swaths overlap by 3 km, the total image swath is 117 km. At the equator, centers of adjacent satellite tracks are separated by a maximum of only 108 km, so in this mode the satellite can acquire completely covered of the Earth's surface.

Off-nadir viewing is possible because the sensor observes the Earth through a pointable mirror that can be controlled by command from the ground. With this capability, the sensors can observe any area within a 950-km swath centered on the satellite track.

When SPOT uses off-nadir viewing, the swath width of individual images varies from 60 to 80 km, depending upon viewing angle. Alternatively, the same region can be viewed from separate positions (from different satellite passes) to acquire stereo coverage. The two sensors are not required to operate in the identical configuration; that is, one HRV can operate in the vertical mode while the other images obliquely. Using its off-nadir viewing capability, SPOT can acquire repeat coverage at intervals of 1 to 5 days, depending upon latitude.

SPOTS 4, 5

SPOT 4, launched in March 1998, and 5, launched in May 2002, continue the SPOT program. A principle feature of the SPOT 4 mission is the high-resolution visible and infrared (HRVIR) instrument, a modification of the HRV used for SPOTS 1, 2 and 3. HRVIR resembles the HRV, with the addition of a mid-infrared band (1.58 ~ 1.75 μm), designed to provide capabilities for geological reconnaissance, for vegetation surveys, and for survey of snow cover.

SPOT 4 carries two identical HRVIR instruments, each with the ability to point 27° to either side of the ground track, providing a capability to acquire data within a 460-km swath for repeat coverage or stereo. In its monospectral (M) mode, HRVIR will provide data in band 2's spectral range at 10-m resolution. In multispectral (X) mode, the HRVIR will acquire four bands of data (1, 2, 3, and mid-infrared) at 20-m resolution.

SPOT 5 carries an upgraded version of the HRVIR, which acquires data at 5-m resolution and provide a capability for along-track stereo imagery. The new instrument, the High-resolution geometrical (HRG), also has the flexibility to acquire data using the same bands and resolutions as the SPOT 4 HRVIR, thereby providing continuity with earlier systems.

Uses of Landsat and SPOT Data

- Geology—is used for mapping in mineral and petroleum exploration
- Agriculture—is used to estimate crop quantities, monitor the condition of the crops
- Forestry—to estimate forest losses caused by fires, clear cutting & disease; to provide forest inventory data; used for comparative forest land valuation
- Land use planning-mapping current land cover, change detection, route location planning
- High resolution satellite imagery is being used as a substitute for high-altitude aerial photography
- For monitoring rangeland condition, wildlife habitat, identify water pollution, identify flooded areas, and to aid in the assessment of damage caused by natural disasters

3.3.3 High Resolution Satellites

1. IKONOS

The IKONOS satellite system, launched in September 1999, is operated by Space Imaging, Denver, Colorado (Private).

In Panchromatic mode, IKONOS provides spatial resolution at 1 meter, in the spectral range

0.45~0.90 mm. In multispectral mode, it provides imagery at 4 m spatial resolution in four spectral bands.

Band 1: 0.45~0.52 μm (blue)
Band 2: 0.52~0.60 μm (green)
Band 3: 0.63~0.69 μm (red)
Band 4: 0.76~0.90 μm (near infrared)

The image swath is 11 km at nadir; imagery is acquired from a sun-synchronous orbit, with a 10:30 a.m. equatorial crossing. The revisit interval varies with latitude; at 40°, repeat coverage can be acquired at about 3 days in the multispectral mode and at about 1 and half days in the panchromatic mode.

2. Quickbird

In October 2001, Earth Watch Inc. (Longmont, Colorado) launched Quickbird, a satellite designed to acquire fine-detailed the imagery using a panchromatic band with detail at 0.61 m resolution and 4 multispectral bands with 2.44 m, details:

Band 1: 0.45~0.52 μm (blue)
Band 2: 0.52~0.60 μm (green)
Band 3: 0.63~0.69 μm (red)
Band 4: 0.76~0.89 μm (near infrared)
Band 5: 0.76~0.89 μm (panchromatic)

Quickbird acquires data using a swath width of 16.5 km.

3. GeoEye 1

GeoEye 1, launched on Sept. 6, 2008, is the world's highest resolution commercial earth-imaging satellite.

GeoEye 1 is equipped with the most sophisticated technology ever used in a commercial satellite system. It offers unprecedented spatial resolution by simultaneously acquiring 0.41-meter panchromatic and 1.65-meter multispectral imagery.

GeoEye's next satellite, GeoEye 2, renamed as WorldView 4, was planned to launch in early 2013 and currently scheduled to launch mid 2016, is in a phased development process for an advanced, third-generation satellite capable of discerning objects on the Earth's surface as small as 0.31-meter in size.

3.4 Mission types

3.4.1 European Space Agency

Since mid-1990s, ESA has developed a dual-mission scenario for Earth observation to address the requirements of the Earth-observation user community: the Earth Watch Missions and the Earth Explorer Missions.

Earth Watch Missions: These are pre-operational missions, each consisting of a series of satellites, addressing the requirements of specific Earth-observation application areas. The responsibility of these types of mission would eventually be transferred to operational (European) entities. The emphasis would be on service. They would address long-term requirements.

Earth Explorer Missions: These would be research/demonstration missions, each of which would focus on advancing the understanding of the processes taking place in the Earth/atmosphere system. The demonstration of specific new observing techniques would also fall into this category. Their duration would vary by mission objectives.

3.4.2 Mission Types—NASA

The mission of NASA's Earth Science Enterprise is to develop a scientific understanding of the Earth system and its response to natural or human-induced changes to enable improved prediction capability for climate, weather, and natural hazards.

The Earth Science Enterprise has defined its research strategy around a hierarchy of scientific questions. At the highest level, the Enterprise is attempting to provide an answer to the one overarching questions:

(1) How is the global system changing ?

(2) What are the primary forcings of the Earth system?

(3) How does the Earth system respond to natural and human-induced changes?

(4) What are the consequences of change in the Earth system for human civilization?

(5) How well can we predict future changes in the Earth system?

These five questions define a pathway of variability, forcing, response, consequence, and prediction that is taken to further enumerate more specific questions that provide direction and focus to the program.

ESE's spaceborne missions fall into three classifications: exploratory, operational precursor & technology demonstration, and systematic measurements.

1. Exploratory Missions

Exploratory missions yield new scientific breakthroughs to promote research and development. Each exploratory satellite project is expected to be a one-time mission that can deliver conclusive scientific results addressing a focused set of scientific questions.

2. Operational Precursor and Technology Demonstration

Requirements for more comprehensive and accurate measurements place increasing pressure on operational environmental agencies and require major upgrades of existing operational observing systems. In order to enable such advances, NASA invests in innovative sensor technologies and develops more cost-effective versions of its pioneer scientific instruments that can be used effectively by operational agencies.

3. Systematic Measurements

Systematic measurements of key environmental variables are essential to specify changes in forcings caused by factors outside the Earth system (e. g. , changes in incident solar radiation) and document the behavior of the major components of the Earth system. The system is not necessarily synonymous with continuous measurement, and gaps in time series may be tolerable when short-term natural variability or calibration uncertainties between discontinuous records do not mask significant long-term trends. ESE aims for continuity in systematic measurements, but does not plan for instantaneous replacement in case of premature sensor or spacecraft failure.

Over the next decade, NASA will transition a number of environmental parameters from research-oriented programs to operationally-oriented ones. The transition presents a challenge requiring careful planning for calibration, retrieval algorithms, and reprocessing of data sets to assure consistency to assure the ability of data from operational entities to address long-term global change questions.

4. Long-term Stable Measurements

AVHRR->MODIS->VIIRS (NPP/NPOESS)

NASA's EOS program was designed to provide long-term stability measurements for climate studies. 3 copies of 3 satellites covering more than 18 year period

Upgraded National Polar-orbiting Operational Environmental Satellite System (NPOESS) program provides similar measurements. Continuing the series of 3 satellites measuring many of the same variables.

Migration of climate community requirements into the operational system is possible, desirable & cost effective.

Initial step is the NPOESS Preparatory Project (NPP). Extends measurement series of 14 parameters begun by MODIS & AIRS.

5. EOS Goals

(1) Develop an understanding of the total Earth system, and the effects of natural and human-induced changes on the global environment.

(2) Expand scientific knowledge of the Earth system using NASA's unique capabilities from the vantage points of space, aircraft, and in situ platforms.

(3) Disseminate information about the Earth system.

(4) Support national and international environmental policy recommendations.

6. Mission Objectives

(1) Create an integrated scientific observing system that will enable multidisciplinary study of Earth system science.

(2) Develop a comprehensive data and information system, including a data retrieval and processing system.

(3) Acquire and assemble a global database emphasizing remote sensing measurements from space over a decade or more.

(4) Improve predictive models of the Earth system.

7. Terra Objectives

(1) Provide the first, consistent global "snapshot" of numerous important Earth surface and

atmospheric characteristics.

(2) Improve the ability to detect the human impacts on climate by identifying indicators, or "fingerprints", of human activity that can be used to distinguish them from natural variability.

(3) Provide measurements of the effects of clouds, aerosols, and greenhouse gases on the Earth's total energy balance.

(4) Provide estimates of global terrestrial and marine productivity that will enable more accurate calculations of global carbon storage, exchange with the atmosphere, and year-to-year variability.

(5) Provide observations that will improve predictions of climate and of weather at seasonal and interannual time scales.

(6) Contribute to improved methods of disaster prediction, characterization, and risk reduction from wild fires, volcanoes, floods, and droughts.

☞ **Summary**

- Platforms
- Data sources with different resolution

☞ **Key-Points**

- Remote sensing platforms
- Orbits and swaths
- Coarse, medium and high resolution satellite data

☞ **Homework**

(1) Make a comparison of Landsat 8 and Landsat 7 spectral bands by logging on http://landsat.usgs.gov, and give a presentation individually in next class.

(2) Investigate major satellite platforms and data sources in China and give a presentation individually in next class.

Chapter 4 Acquiring Remote Sensing Data

Remote sensors can be grouped according to the number of bands and the frequency range of those bands that the sensor can detect. Common categories of remote sensors include panchromatic, multispectral, hyperspectral, and ultraspectral sensors.

Panchromatic sensors cover a wide band of wavelengths in the visible or near-infrared spectrum. An example of a single band sensor of this type would be a black and white photographic film camera.

Multispectral sensors cover two or more spectral bands simultaneously typically from 0.3 m to 14 m wide.

Hyperspectral sensors cover spectral bands narrower than multispectral sensors. Image data from several hundred bands are recorded at the same time offering much greater spectral resolution than a sensor covering wider bands(Figure 4.1).

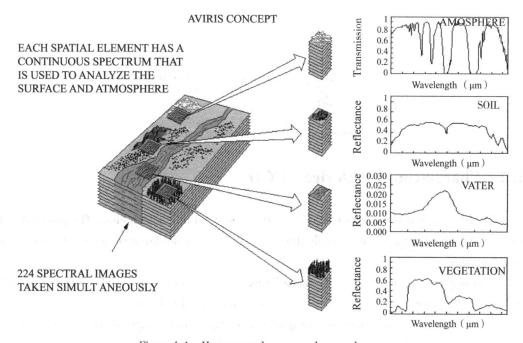

Figure 4.1 Hyperspectral sensor and spectral curves

Ultrasprectral sensors are still under development and not yet in use. These sensors will cover thousands of bands with an even narrower bandwidth than hyperspectral sensors.

Digital images can be generated by two kinds of instruments, optical-mechanical scanners and charge-coupled device (CCDs).

4.1 Optical-mechanical scanners

Optical-mechanical scanners physically move mirrors or lenses to systematically aim the field of view over the Earth's surface. As the instrument scans the Earth's surface, it generates an electrical current that varies in intensity as the land surface varies in brightness (Figure 4.2).

Sensors sensitive in several regions of the spectrum use filters to separate energy into several spectral regions, each represented by a separate electrical current. Each electrical signal must be subdivided into distinct units to create the discrete values necessary for digital analysis. This conversion from the continuously varying analog signal to the discrete digital values is accomplished by sampling the current at a uniform interval (analog-to-digital, or A- to-D, conversion) (Figure 4.3).

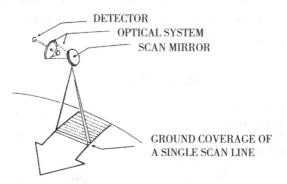

Figure 4.2 Optical-mechanical scanner

4.2 Charge-coupled Device (CCD)

A CCD is formed from light-sensitive material embedded in a silicon chip. The potential well receives photons from the scene, usually through an optical system designed to collect, filter, and focus radiation.

Then sensitive components of CCDs can be manufactured to be very small, perhaps as small as 1 μm in diameter, and sensitive to visible and near-infrared radiation.

These elements can be connected using microcircuitry to form arrays; detectors arranged in a single line form a linear array, detectors arranged in several rows and columns from two-dimensional arrays.

CCDs can be positioned in the focal plane of a sensor such that they view a thin rectangular

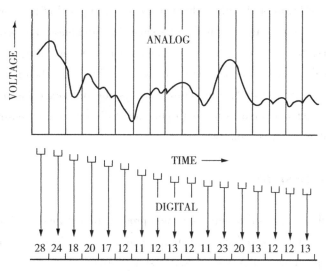

Figure 4.3 Analog to digital conversion (A-D)

strip oriented at right angles to the flight path. The forward motion of the aircraft or satellite moves the field of view forward along the flight path, building up coverage.

By analogy, this means of generating an image is known as push-broom scanning. In contrast, mechanical scanning can be visualized by analogy to a whiskbroom, which creates an image using the side-to-side motion of the scanner.

S/N Ratio

Each sensor creates responses unrelated to target brightness—that is, noise, create in part by accumulated electronic errors from various components of the sensor. For effective use, the instruments must be designed such that their noise levels are small relative to the signal. This is measured as the signal-to-noise ratio (S/N or SNR) (Figure 4.4).

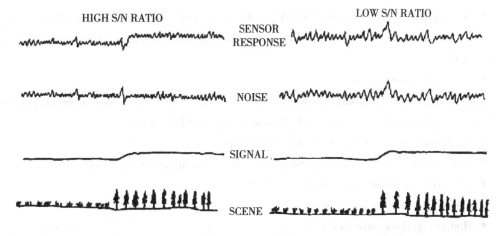

Figure 4.4 S/N Ratio

At the bottom of Figure 4.4, a hypothetical scene is composed of two cover types. The signal records this region, with only a small difference in brightness between the two classes. Atmospheric effects, sensor error, and other factors contribute to noise, which is added to the signal. The sensor then records a combination of signal and noise. When noise is small relative to the signal (left: high S/N ratio), the sensor conveys the difference between the two regions. When the signal is small relative to noise (right: low S/N ratio) the sensor cannot portray the difference in brightness between the two regions.

4.3 Scanner Systems

Electro-optical and spectral imaging scanners produce digital images with the use of detectors that measure the brightness of reflected electromagnetic energy. Scanners consist of one or more sensor detectors depending on the type of sensor system used.

4.3.1 Whiskbroom Scanner

One type of scanner is called a whiskbroom scanner also referred to as a cross-track scanners. It uses rotating mirrors to scan the landscape below from side to side perpendicular to the direction of the sensor platform, like a whisk broom (Figure 4.5). The width of the sweep is referred to as the sensor swath. The rotating mirrors redirect the reflected light to a point where a single or just a few sensor detectors are grouped together. Whiskbroom scanners with their moving mirrors tend to be large and complex to build. The moving mirrors create spatial distortions that must be corrected with preprocessing by the data provider before the image data is delivered to the user.

- Such systems scan the terrain along scan lines that are at right angles to the flight line. This allows the scanner to repeatedly measure the energy from one side of the aircraft to the other
- Data are collected within an arc of 90°~120°
- Instantaneous field of view (IFOV): the cone angle within which incident energy is focused on the detector. IFOV is determined by the optical system and size of the detectors (Figure 4.6)
- Pure and mixed pixels.

Figure 4.6 shows that ground resolution cells are larger towards the edge of the image. Small IFOV means high spatial detail recorded, however, larger IFOV means:

- Greater quantity of total energy on a detector
- More sensitive scene radiance measurements due to higher signal levels
- Improved radiometric resolution
- Signal greater than background noise
- Higher signal-to-noise ration
- Longer dwell time

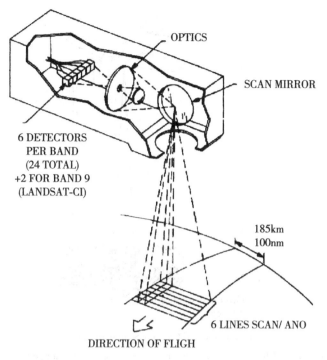

Figure 4.5 Whiskbroom scanning

4.3.2 Push-broom (Along-track)

Another type of scanner, which does not use rotating mirrors, is the push-broom scanner also referred to as an along-track scanner. The sensor detectors in a push-broom scanner are lined up in a row called a linear array. Instead of sweeping from side to side as the sensor system moves forward, the one dimensional sensor array captures the entire scan line at once like a push-broom would (Figure 4.7).

Some recent scanners referred to as step stares scanners contain two-dimensional arrays in rows and columns for each band.

Push-broom scanners are lighter, smaller and less complex because of fewer moving parts than whiskbroom scanners. Also, they have better radiometric and spatial resolution.

A major disadvantage of push-broom scanners is the calibration required for a large number of detectors that make up the sensor system.

The size of detectors determines the size of each ground resolution cell.

Each spectral band (or channel) requires its own array.

Advantages of Push-broom over Whiskbroom

(1) Longer dwell time, stronger signal, greater range of sensed signal, better spatial and radiometric resolution.

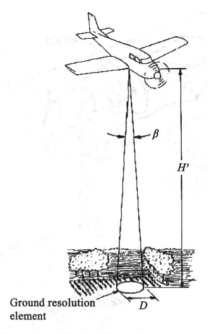

Figure 4.6 IFOV and Spatial resolution

D: diameter of the circular ground area viewed (spatial resolution)

β: Instantaneous field of view

H': Flying height above terrain $D = H'\beta$

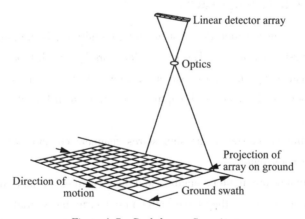

Figure 4.7 Push-broom Scanning

(2) Better geometry (fixed relationship among detector elements).

(3) Lighter and smaller devices, require less energy.

Disadvantages of Push-broom over Whiskbroom

(1) Need to calibrate more detectors.

(2) Limited range of spectral sensitivity of commercially available CCDs.

4.4 Remotely Sensed Data

4.4.1 Raster or Grid Format

Pixel is the smallest areal units identifiable on the image.

Features with regular shapes, such as squares, rectangles or hexagons. Squares are the preferred shapes.

Imaginary matrix (row & column format) is placed on the feature (e.g., the ground). Some phenomena (e.g., the amount of reflected light) is measured. A value representing the strength of the signal is assigned to each grid cell (pixel).

For remote sensing, the average amount of EMR from the area of a pixel on the ground is "measured."

Depending on the average intensity of the EMR a numeric value is assigned to each pixel. In which, low or none radiance is the minimum value, high radiance is the maximum value, and others are scaled between the minimum and maximum values.

4.4.2 Digital Image Data

- Digital data are matrices of digital numbers (DNs) (Figure 4.8)
- There is one layer (or matrix) for each satellite band
- Each DN corresponds to one pixel
- Images are presented as 2-d arrays. Each pixel (array element) has a location (x, y)

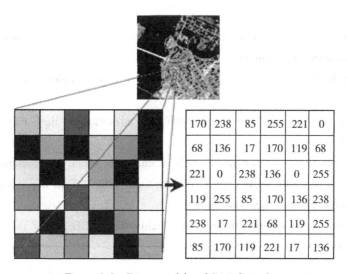

Figure 4.8 Raster model and Digital number

and associated with it a digital number (DN)
- Position of pixel often describes in terms of rows and columns

1. What are digital numbers (DNs)

DNs are relative measures of radiance, and DNs are not reflectance. DNs can be converted to ground reflectance, if you know atmospheric properties, etc. The range of DNs depends on the radiometric resolution of the instrument.

2. Image File Formats

Many proprietary image file formats, typically include a header and the image data themselves.

Image data can be organized in several ways, such as Band Sequential Formal (BSQ), Band Interleaved by Line (BIL) and Band Interleaved by Pixel (BIP).

4.4.3 Data Acquisition

The detection/recording of reflected/emitted EMR from the objects can be performed either photographically or electronically.

The photographic film acts as both the detecting and recording medium.

Electronic sensors generate an electrical signal that corresponds to the energy variation in the original scene.

Subsequently the signals can be converted to an image or photographic film.

4.4.4 Data Interpretation

Data interpretation aspects of remote sensing can involve analysis of image data.

Visual interpretation of pictorial image data has long been the most common form of remote sensing.

Visual techniques make use of the excellent ability of the human mind to qualitatively evaluate spatial patterns in an image.

Elements such as shape, size, pattern, tone, texture, shadow, site, association, and resolution.

Visual interpretation techniques have certain disadvantages (they may require extensive training, labor intensive).

Spectral characteristics are not always fully evaluated in visual interpretation.

Due to the limited ability of the eye to discern tonal values of an image and the difficulty of simultaneously analyzing multi spectral images.

In applications where spectral patterns are highly informative, it is therefore preferable to analyze digital, rather than pictorial image data.

4.4.5 Reference Data

Remote sensing is always employed with some forms of reference data.

The acquisition of reference data (ground truth) involves collecting measurements or observations about the objects, areas, or phenomena that are being sensed remotely. For example, the data needed for a particular analysis might be derived from a soil survey map, a water quality laboratory report, or an aerial photograph, etc.

They may also stem from the field check, on the identity, extent and condition of agricultural crops, land use, tree species, or water pollution problem.

Reference data might be used to serve any or all of the following purposes: (1) to aid in the analysis and interpretation of remotely sensed data; (2) to calibrate a sensor and (3) to verify information extracted from remote sensing data.

☞ **Summary**

- Optical-mechanical scanners and CCD
- Whiskbroom vs. push-broom scanning
- Digital data (Digital Data Format and Storage)

☞ **Key Points**

- Optical-mechanical scanners and charge-coupled device (CCDs)
- S/N Ratio
- Whiskbroom scanner and push-broom scanner
- Raster data model and image file format
- DN values

☞ **Homework**

Investigation softwares for processing remotely sensed images, and give a presentation in next class individually.

Chapter 5 Image corrections and preprocessing

5.1 Preprocessing

Preprocessing refers to those operations that are preliminary to the main analysis. Typically, preprocessing operations could include (1) geometric preprocessing to bring an image into registration within a map or another image, and (2) radiometric preprocessing to adjust digital values for the effect of a hazy atmosphere.

Once corrections have been made, the data can then be subjected to the primary analysis.

Preprocessing forms a preparatory phase that, in principle, improves image quality as the basis for later analyses.

Preprocessing includes a wide range of operations. Most can be categorized into one of the three groups: (1) feature extraction, (2) radiometric correction, and (3) geometric corrections.

5.2 Geometric Corrections

1. What is Geometric Correction?

Any process that changes the spatial characteristics of the pixels. Such as the change of pixel coordinates, the pixel relationship with other pixels and the alteration of pixel size.

Geometric correction also can change the radiometry of pixels (re-sampling).

2. Why Geometric Correction?

(1) To allow an image to overlay a map.

(2) To warp an image to eliminate distortion caused by terrain, instrument wobble, earth curvature, etc.

(3) To change the spatial resolution of an image.

(4) To change the map projection used to display an image.

3. Two basic strategies for fitting images to map

(1) Use Ground Control Points (GCPs) to assign real-world coordinates to an image (rectification).

(2) Create links between two images or between an image and a digital map to line them up with one another (registration).

Both strategies based on the same concept.

5.2.1 Rectification Using Cntrol Points

1. Object

To match pixel locations in the image to their corresponding locations on the Earth (Figure 5.1).

2. Method

(1) Assign coordinates to known locations in the image (Ground Control Points = GCPs).

(2) Create a "model" to fit all GCPs.

(3) "Warp" image to best fit the model.

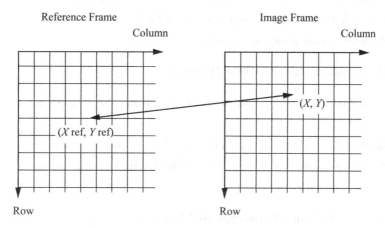

Figure 5.1　Image rectification

3. Identification of Ground Control Points (GCPs)

GCPs are features that can be located with precision and accuracy on accurate maps yet are also easily located on digital images.

Ideally, GCPs could be as small as a single pixel, if one could be easily identified against its background.

In practice, most GCPs are likely to be spectrally distinct areas as small as a few pixels.

It might include intersections of major highways, distinctive water bodies, edges of land-cover parcels, stream junctions, and similar features.

Typically, it is relatively easy to find a rather small or modest-sized set of control points. However, in some scenes, the analyst finds it increasingly difficult to expand this set, as one has less and less confidence in each new point added to the set of GCPs.

Thus, there may be a rather small set of "good" GCPs, points that the analyst can locate with confidence and precision both on the image and on an accurate map of the region.

The locations may also be a problem. In principle, GCPs should be dispersed throughout the image, with good coverage near edges.

The desire to select "good" GCPs and to achieve good dispersion may work against each other, such that the analyst finds it difficult to select a judicious balance.

Registration error decreases as the number of GCPs is increased (Bernstein, et al., 1983). It is better to have more rather than fewer GCPs.

It is recommended that 16 GCPs may be a reasonable number if each can be located with an accuracy of one-third of a pixel. This number may not be sufficient if the GCPs are poorly distributed or if the nature of the landscape prevents accurate placement.

(1) Road intersections, river bends, distinct natural features, etc.

(2) GCPs should be spread across the image.

(3) Requires a minimum number depending on the type of transformation.

(4) Some say that it is better to have clusters of GCPs.

(5) Must choose a map projection for GCP coordinates.

5.2.2　How is image registration different?

- Instead of finding ground control points from a map, you link the same place on two or more images
- Can be used to georeference an unreferenced image using a referenced image
- Can be used to allow two images to perfectly line up with one another (e.g., images of the same place taken on different dates)

5.2.3　Processes that allow images to fit a map

Transformation: Use a mathematical equation to transform pixel coordinates to best match the coordinates of all of the GCPs simultaneously.

Resampling: Assign DNs to the pixels once they have been moved to their new positions.

1. Mathematical Transformations

A mathematical transformation is calculated by regressing the original x/y coordinates against the georectified coordinates for the same points. This transformation is then applied to all the data. Images can be transformed using 1st, 2nd, and 3rd order polynomial transformations. A 1st-order transformation is a linear transformation. It requires minimum of 3 GCPs, and generally used for small and flat areas. Transformations of the 2nd-order, 3rd-order or higher are nonlinear transformations. The 2nd-order transformation requires minimum of 6 GCPs, and generally used for larger area where earth curvature is a factor or a moderate terrain. And the 3rd-order transformation requires minimum of 10 GCPs, and generally used for areas with very rugged terrain (Figure 5.2).

Typically, want at least 3x the minimum number of GCPs.

2. Image Resampling

Once an image is warped, how do you assign DNs to the "new" pixels?

Resampling Techniques include:

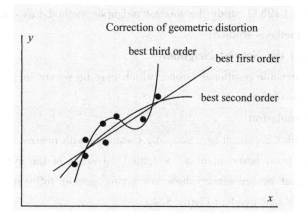

Figure 5.2　Mathematical transformations for correction
of geometric distortion

(1) Nearest Neighbor: Assign the value of the nearest pixel to the new pixel location.

(2) Bilinear: Assign the average value of the 4 nearest pixels to the new pixel location.

(3) Cubic Convolution: Assign the average value of the 16 nearest pixels to the new pixel location.

3. Nearest Neighbor

Assign pixel the value of each "corrected" from the nearest "uncorrected" pixel (Figure 5.3)

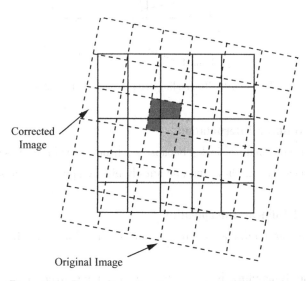

Figure 5.3　Resampling using the nearest neighbor technique

Advantages of the Nearest Neighbor

(1) It has the advantages of simplicity and the ability to preserve the original values in the unaltered scene.

(2) In Kovalick's (1983) study the nearest neighbor method was computationally the most efficient of the three methods studied.

Disadvantages of the Nearest Neighbor

It may create noticeable positional errors, which may be severe in linear features where the realignment of pixels is obvious.

4. Bilinear Interpolation

It is a more complex approach to resampling compared with nearest neighbor. It calculates a value for each output pixel based upon a "weighted" average of the four nearest input pixels. "Weighted" means that nearer pixel values are giving greater influence in calculating output values than are more distant pixels. (Figure 5.4)

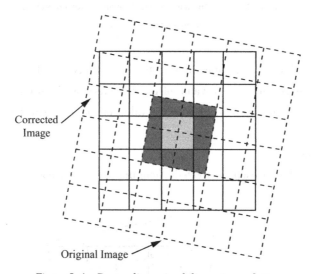

Figure 5.4 Resampling using bilinear interpolation

Advantages of Bilinear Interpolation

The image has a more "nature" look: because each output value is based upon several input values, the output image will not have the unnaturally blocky appearance of some nearest neighbor images.

Disadvantages of Bilinear Interpolation

(1) Because bilinear interpolation creates new pixel values, the brightness values in the input image are lost.

(2) Because the resampling is conducted by averaging over areas, it decreases spatial resolution by a kind of "smearing" caused by averaging small features with adjacent background pixel.

5. Cubic Convolution

It is the most sophisticated, most complex, and (possibly) most widely used resampling method.

Cubic convolution uses a weighted average of values within a neighborhood that extents about two pixels in each direction, usually encompassing 1 adjacent pixels (Figure 5.5).

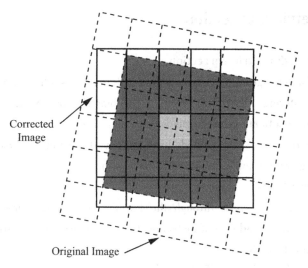

Figure 5.5 Resampling using cubic convolution

Advantages of Cubic Convolution

The images produced by cubic convolution resampling are much more attractive than those of other procedures for cubic convolution has higher precision.

Disadvantages of Cubic Convolution

(1) The data are altered more than those of nearest neighbor or bilinear interpolation.

(2) The computation is more intensive, and the minimum number of GCPs is larger.

To maintain image radiometry for spectral analysis always use the nearest neighbor resampling!

If your purpose is to produce a smoothed image for presentation, bilinear or cubic might work better.

Remember that every time you resample an image for any reason you are altering the original data.

6. Changing the Spatial Resolution (A type of Resampling)

(1) Increase the resolution (artificially make the pixels smaller): just assign the DN from the original pixel to the smaller pixels that fall inside it.

(2) Decrease the resolution (artificially make the pixels larger): combine the DNs from the original pixels in some way (e.g., average them) to assign a new DN to the bigger pixel.

5.2.4 Summary of Geometric Correction

- Essential for almost all remote sensing projects

- Allows images to correspond to real world map coordinates
- Critical for combining images and GIS
- Essential for obtaining spatially accurate products—requires considerable care

5.3 Radiometric Correction

1. What is the radiometric correction

Any process that changes the DNs of individual pixels, such as atmospheric corrections, contrast stretching, filtering, destriping, etc., is radiometric correction.

2. What are atmospheric corrections

Atmospheric "corrections" are methods used to convert the radiance measured at the satellite to reflectance or outgoing radiance measured at the ground.

3. Atmospheric Effects

EMR from the sun must pass through the Earth's atmosphere twice before it reaches the satellite. The atmosphere is made up of molecules that can interact with EMR. Often, atmospheric effects are wavelength dependent.

There are 2 (or 3) primary effects of the atmosphere on EMR, mainly are atmospheric scattering, atmospheric absorption (or the opposite, transmission/transmittance). EMR can also be refracted (bent) by the atmosphere.

5.3.1 Introduction to Atmospheric Correction

Any sensor that observes the Earth's surface using visible or near infrared radiation will record a mixture of two kinds of brightnesses. One brightness is due to the reflectance from the Earth's surface—the brightness that is of interest in remote sensing. But the sensor also observes the brightness of the atmosphere itself—the effect of scattering.

Thus, an observed digital brightness value might be in part the result of surface reflectance and in part the result of atmospheric scattering. We cannot immediately distinguish the two brightnesses, so one objective of atmospheric correction is to identify and separate these two components so the main analysis can focus upon examination of correct surface brightness.

Most atmospheric corrections attempt to remove path radiance (L_p) by subtracting it out. Assuming similar atmospheric conditions over the entire image (treat every pixel the same), this must be done for each band separately. Because green radiation is scattered by atmospheric 4x more than near infrared, and atmospheric effects are much stronger in visible part of the spectrum in more general than in the IR.

5.3.2 Correcting for Path Radiance

This must be done separately for each band, and must estimate amount of path radiance and remove it (subtract) from each band. Techniques included "dark pixel subtraction" or regression of short wavelength band against long (unscattered) wavelength bands.

1. Dark Pixel Subtraction

Assumes that the darkest objects in the image should have a DN of 0 (no reflectance), of course, this is not always a correct assumption.

Find the minimum pixel value from each band (using histograms) (Figure 5.6).

Subtract that value from all of the pixels in the band.

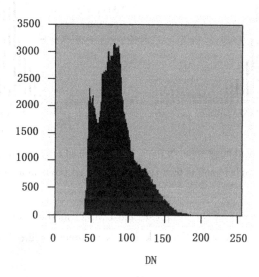

Figure 5.6 Histogram shows the minimum DN

2. Dark Pixel Subtraction—A Case Study

In its simplest form, this strategy can be implemented by identifying a very dark object or feature within the scene. Such an object might be a large water body or possibly shadows cast by clouds or by large topographic features.

In the infrared portion of the spectrum, both water body clouds and shadows should have brightness at or very near zero because clear water absorbs strongly in the near infrared spectrum and because very little infrared energy is scattered to the sensor from shadowed pixels.

Analysts who examine such areas, or the histograms of the digital values for a scene, can observe that the lowest values (for dark area, such as water bodies) are not zero, but some large values.

Typically, this value will differ from one band to the next, so, for example, for Landsat band 1 the value might be 12, for band 2 the value is 7, for band 3 the value is 2, and for band 4 the value is 2. These values, assumed to represent the value contributed by atmospheric scattering for each band, are then subtracted from all digital values for that scene and that band (Figure 5.7).

Thus, the lowest value in each band is set to zero, the dark black color assumed to be the correct tone for a dark object in the absence of atmospheric scattering.

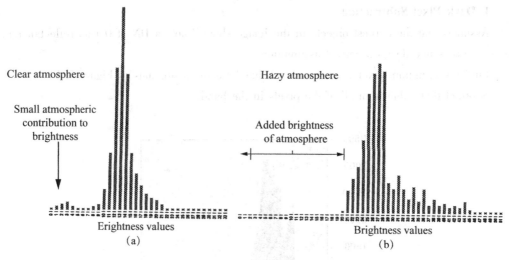

Figure 5.7 The lowest brightness value in the scene is assumed to reveal the added brightness of the atmosphere and is then subtracted from all pixels in a specific band

This procedure forms one of the simplest, most direct methods for adjusting digital values for atmospheric degradation (Chavez, 1975), known sometimes as the histogram minimum method (HMM).

This procedure has the advantages of simplicity, directness, and almost universal applicability, as it exploits information present within the image itself.

Yet, it must be considered as an approximation; atmospheric effects change not only the position of the histogram of the axis, but also its shape (i.e., not all brightnesses are affected equally). In addition, in arid regions observed at high sun angles, shadows, clouds, and open water may be so rare or no, such small area extent that the method cannot be applied.

5.3.3 Correcting for Atmospheric Transmittance

- Must correct separately for each band
- Must either measure or make assumptions about optical depth, atmospheric density of various constituents, etc.
- Often not done because it is complicated and difficult

5.3.4 Atmospheric Measurement and Modeling

Require measurement of many atmospheric characteristics at different heights above the Earth at the same time as satellite overpass.

There are "canned" atmospheric models that work fairly well, such as LOWTRAN, MODTRAN.

LOWTRAN calculates atmospheric transmittance and atmospheric background radiance for a variety of atmospheric condition.

LOWTRAN applies relatively low spectral resolution, whereas MODTRAN uses a somewhat finer spectral resolution.

These programs estimate atmospheric absorption and emission of atmospheric gasses.

These models accommodate differing atmospheric conditions, including seasonal and geographic variations, cloud conditions, rain, and haze. These models consider all possible atmospheric paths.

5.3.5 Reasons for Atmospheric Correction

- Atmospheric Correction is not always necessary

(1) Single scene studies.

(2) Atmospheric differences can be reduced by ratio based vegetation indices (like NDVI).

- Often necessary when comparing multiple scenes

(1) Scene matching (mosaics).

(2) Change detection studies.

(3) Applying classification statistics to multiple scenes.

- Always necessary if you need to calculate ground reflectance or compare satellite to ground reflectance measurements.

5.3.6 Summary of Radiometric Corrections

- Change the DNs of pixels from the values that the satellite measured
- Usually done to remove radiance not directly from the target (e.g., path radiance)
- Should be considered carefully because you alter the original data

5.4 Feature Extraction

In the context of image processing, the term feature extraction (or feature selection) has specialized meaning. "Features" are not geographical feature, visible in an image, but are rather "statistical" characteristics of image data: individual bands or combinations of band values that carry information concerning systematic variation within the scene. Thus, feature extraction could also be known as "information extraction", isolation of components within multispectral data that are most useful in portraying the essential elements of an image. In theory, discarded data contain noise and errors present in the original data. Thus feature extraction may increase accuracy. In addition, feature extraction reduces the number of spectral channels, or bands that must be analyzed, thereby reducing computational demands. After feature selection is complete, the analyst works with fewer but more potent channels. The reduced data set may convey almost as much information as does the complete data set. Feature selection may increase speed and reduce costs of analysis.

Multispectral data, by their nature, consists of several channels of data. Although some images may have as few as 3, 4, or 7 channels, other image data may have many more, possibly 200 or more channels. With so much data, processing of even modest-sized images requires considerable time. In this context, feature selection assumes considerable practical significance, as image analysts wish to reduce the amount of data while retaining effectiveness and/or accuracy.

Our examples here are based on TM data, which provide enough channels to illustrate the concept, but are compact enough to be reasonably concise. A variant-covariance matrix shows the interrelationships between pairs of bands; some pairs show rather strong correlations—for example, bands 1 and 3, bands 2 and 3 both show correlations above 0.9. High correlation between pairs of bands means that the values in the two channels are closely related. Thus, as values in channel 2 rise or fall, so do those in channel 3; one channel tends to duplicate information in the other. Feature selection attempts to identify, and then remove, such duplication so that the data set can include maximum information using the minimum number of channels.

For example, for the data represented by Table 5.1, bands 3, 5, and 6 might include almost as much information as the entire set of seven channels because band 3 is closely related to bands 1 and 2, band 5 is closely related to bands 4 and 7, and band 6 carries information largely unrelated to any others. Therefore, the discarded channels (1, 2, 4 and 7) each resemble one of the channels that have been retained. So a simple approach to feature selection discards unneeded bands, thereby reducing the number of channels. Although this kind of selection can be used as a kind of rudimentary feature extraction, typically, feature selection is a more complex process based upon statistical interrelationships between channels.

A common approach to feature selection applies a method of data analysis called principal component analysis (PCA) (Davis, 1986). This text will offer only a superficial description, as more complete explanation requires the level of detail provided by Davis (1986) and others. In essence, PCA identifies the optimum linear combinations of the original channels that can account for variation of pixel values in an image. Linear combinations are of the form

$$A = C_1X_1 + C_2X_2 + C_3X_3 + C_4X_4$$

Where, X_1, X_2, X_3, and X_4 are pixel values in four spectral channels, and C_1, C_2, C_3, and C_4 are coefficients applied individually to the values in the respective channels. A represents a transformed value for the pixel. Assume, as an example, that $C_1 = 0.35$, $C_2 = -0.08$, $C_3 = 0.36$, and $C_4 = 0.86$. For a pixel with $X_1 = 28$, $X_2 = 29$, $X_3 = 21$, $X_4 = 54$, the transformation assumes a value of 61.48. Optimum values for coefficients are calculated by a procedure that ensures that the values they produce account for maximum variation within the entire data set. Thus, this set of coefficients provides the maximum information that can be conveyed by any single channel formed by a linear combination of the original channels. If we make an image from all the values formed by applying this procedure to an entire image, we generate a single band of data that provides an optimum depiction of the information present within the four channels of the original scene.

Table 5.1   Similarity matrices for seven bands of TM scene

Covariance matrix

	1	2	3	4	5	6	7
1	48.8	29.2	43.2	49.9	76.5	0.9	44.9
2	29.2	20.3	29.0	48.6	65.4	1.5	32.8
3	43.2	29.0	46.4	59.9	101.2	0.6	53.5
4	49.9	48.6	59.9	327.8	325.6	12.4	104.3
5	76.5	65.4	101.2	325.6	480.5	10.2	188.5
6	0.9	1.5	0.6	12.5	10.2	14.0	1.1
7	45.0	32.8	53.5	104.3	188.5	1.1	90.8

Correlation matrix

	1	2	3	4	5	6	7
1	1.00						
2	0.92	1.00					
3	0.90	0.94	1.00				
4	0.39	0.59	0.48	1.00			
5	0.49	0.66	0.67	0.82	1.00		
6	0.03	0.08	0.02	0.18	0.12	1.00	
7	0.67	0.76	0.82	0.60	0.90	0.02	1.00

The effectiveness of this procedure depends, of course, upon calculation of the optimum coefficients. Here our description must be, by intention, abbreviated because calculation of the coefficients is accomplished by methods requiring full explanations, such as those given by upper level statistics texts or discussions such as those of Davis (1986), and Gould (1967). For the present, the important point is that PCA permits identification of a set of coefficients that concentrates maximum information in a single band.

The same procedure also yields a second set of coefficients that will yield a second set of values (we could represent this as the B set, or B image) that will be a less effective conveyor of information, but will represent variations of pixels within the image. In all, the procedure will produce seven sets of coefficients (one set for each band in the original image), and therefore will produce seven sets of values, or bands (here denoted as A, B, C, D, E, E, and G), each in sequence conveying less information than the preceding band. Thus, in Table 5.2 transformed channels I and II (each formed from linear combinations of the seven original channels) together account for only about 93% percent of the total variation in the data, whereas channels III ~ VII together account for only about 7% of the total variance. The analysts may be willing to discard the variables that convey 7% of the variance as a means of reducing the number of channels. The analysts still retain 93% of the original information in a much more concise form. Thus, feature selection reduces the size of the data set by eliminating replication of information.

Table 5.2   **Results of principal components analysis of Data in Table 5.1**

	Components						
	I	II	III	IV	V	VI	VII
	Eigenvectors						
%Var	82.5%	10.2%	5.3%	1.3%	0.4%	0.3%	0.1%
EV	848.44	104.72	54.72	13.55	4.05	2.78	0.77
	0.14	0.35	0.60	0.07	−0.14	−0.66	−0.20
	0.11	0.16	0.32	0.03	−0.07	−0.15	−0.90
	0.37	0.35	0.39	−0.04	−0.22	0.71	−0.36
	0.56	−0.71	0.37	−0.09	−0.18	0.03	−0.64
	0.74	0.21	−0.50	0.06	−0.39	−0.10	0.03
	0.01	−0.05	0.02	0.99	0.12	0.08	−0.04
	0.29	0.42	−0.08	−0.09	0.85	0.02	−0.02
	Loadings						
Band 1	0.562	0.519	0.629	0.037	−0.040	−0.160	−0.245
Band 2	0.729	0.369	0.529	0.027	−0.307	−0.576	−0.177
Band 3	0.707	0.528	0.419	−0.022	−0.659	−0.179	−0.046
Band 4	0.903	−0.401	0.150	−0.017	0.020	0.003	−0.003
Band 5	0.980	0.098	−0.166	0.011	−0.035	−0.008	−0.001
Band 6	0.144	−0.150	0.039	0.969	0.063	0.038	−0.010
Band 7	0.873	0.448	−0.062	−0.033	0.180	0.004	−0.002

The effect is easily seen in Figure 5.8, which shows transformed data for a subset of a Landsat TM scene. Images PC I and PC II are the most potent; PC III, PC IV, PC VI, and PC VII show the decline in information content, such that the final images record (one assume) artifacts of system noise, atmospheric scattering, and other undesirable contributions to image brightness. If these two channels are excluded from subsequent analysis, it is likely that the accuracy can be retained (relative to the entire set of four channels) while also reducing costs.

This method is not the only means for feature selection, but is does illustrate the objectives of this step: to reduce the number of channels to be examined while simultaneously retaining as much information as possible and reducing the contributions of noise and error.

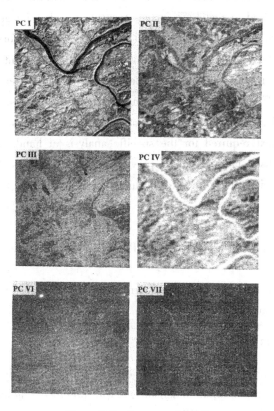

Figure 5.8 Feature selection

These images depict six of the seven principal components of the image described by Tables 5.1 and 5.2. The first principal component image (PC I), formed from a linear combination of data from all seven original bands, accounts for 82.5% of the total variation of the image data. PC II and PC III present 10.2% and 5.3% of the total variation, respectively. The higher components (i.e., PC VI and PC VII) account for very low proportions of the total variation, and convey mainly noise and error, as is clear by the image patterns they show

5.5 Subsets

Because of the very large sizes of many remotely sensed images, analysts typically work with those segments of full image that specifically pertain to the task at hand. Therefore, to minimize computer storage, and the analyst's time and effort, one of the first tasks in each project is to prepare subsets, portions of larger images selected to show only the region of interest.

Although selecting subsets would not appear to be one of remote sensing's most challenging tasks, it is more difficult than one might first suppose. Often subsets must be "registered" (matched) to other datas, or to other projects, so it is necessary to find distinctive landmarks in both sets of datas to assure that coverages coincide spatially. Second, since time and computational effort devoted to matching images to maps or other images (as described below) increase with large images, it is often convenient to prepare subsets before registration. Yet, if the subset is too small, then it may be difficult to identify sufficient landmarks for efficient registration. Therefore, it may be useful to prepare a preliminary subset, large enough to conduct the image registration effectively, before selecting the final, smaller, subset for analytical use.

The same kinds of considerations apply in other steps of an analysis. Subsets should be large enough to provide context required for the specific analysis at hand. For example, it may be important to prepare subsets large enough to provide sufficient numbers of training fields for image classification, or a sufficient set of sites for accuracy assessment.

☞ **Key Points**

- Geometric correction: Ground Control Points (GCPs)/Image Matching; Resampling Techniques
- Radiometric correction: atmospheric corrections (dark pixel subtraction)
- Principal components analysis

☞ **Review Questions**

(1) How can an analyst determine if specific preprocessing procedures have been effective?

(2) Suppose an enterprise offers to sell images with preprocessing already completed. Would such a product be attractive to you? Why or why not?

Chapter 6 Image Interpretation

6.1 Introduction

Such information is not presented to us directly: the information we seek is encoded in the varied tones and textures we see in each image.

To translate images into information, we must apply a specialized knowledge—knowledge that forms the field of image interpretation, which we can apply to derive useful information from the raw uninterpreted images we receive from remote sensing systems.

6.1.1 Needed for Image Interpretation

Subject: the knowledge of the subject of our interpretation—the kind of information that motivates us to examine the image—is the heart of the interpretation;

Geographical region: the unique characteristics that influence the patterns recorded on an image;

Remote sensing system: the variables that influence the image to be interpreted and how to evaluate them.

6.1.2 Image Interpretation Tasks

1. Image Interpretation Tasks—Classification

Classification is the assignment of objects, features, or areas to classes based on their appearance on the imagery.

Detection: the determination of the presence or absence of a feature.

Recognition: a higher level of knowledge about a feature or object, such that the object can be assigned an identity in a general class or category.

Identification: the identity of an object or feature can be specified with enough confidence and detail to place it in a very specific class (Figure 6.1).

Often, an interpreter may qualify his or her confidence in an interpretation by specifying the identification as "possible" or "probable".

2. Image Interpretation Tasks—Enumeration

Enumeration is the task of listing or counting discrete items visible in an image (Figure 6.2).

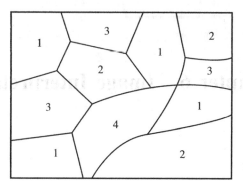

Figure 6.1 Classification

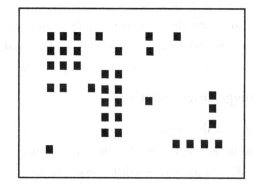

Figure 6.2 Enumeration

3. Image Interpretation Tasks—Measurement

First is the measurement of distance and height, and, by extension, of volumes and areas as well.

A second form of measurement is quantitative assessment of image brightness (Figure 6.3).

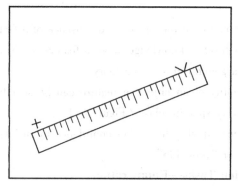

Figure 6.3 Measurement

4. Image Interpretation Tasks—Delineation

Separate distinct areal units that are characterized by specific tones and textures, and to identify edges or boundaries between separate areas.

6.2　Elements of Image Interpretation

The main elements of image interpretation are as follows:
- Tone
- Texture
- Shadow
- Pattern
- Association
- Shape
- Size
- Site

6.2.1　Tone

Image tone denotes the lightness or darkness of a region within an image.

For black-and-white image, tones may be characterized as light, medium gray, dark gray, dark and so on, as the image assumes varied shades of white, gray, or black (Figure 6.4).

For color images, image tone refers simply to "color".

Interpreters tend to be consistent in interpretation of tones on black-and-white imagery, but less so in interpretation of color imagery. A human interpreter can provide reliable estimates of relative differences in tone, although he or she may not be capable of accurate description of absolute image brightness.

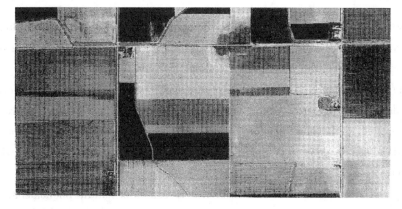

Figure 6.4　Image tone

6.2.2 Texture

Image texture is defined as the spatial variation of image tone. It may be described in many ways such as smooth, rippled, lineated, irregular, and patchy.

Usually, texture is caused by the pattern of highlighted and shadowed areas created when an irregular surface is illuminated from an oblique angle (Figure 6.5).

The human interpreter is very good at distinguishing subtle differences in image texture, so it is a valuable aid to interpretation—certainly equal in importance to image tone in many circumstances.

Figure 6.5 Image texture

6.2.3 Shadow

Shadow is an especially important clue in the interpretation of objects. A building or vehicle, illuminated at an angle, casts a shadow that may reveal characteristics of its size or shape that would not be obvious from the overhead view alone.

The role of shadow in the interpretation of any man-made landscape in which identification of separate kinds of structures or objects is significant. Because military photo interpreters often are primarily interested in the identification of individual items of equipment, they have developed many methods to use shadows to distinguish subtle differences that might not otherwise be visible (Figure 6.6).

Shadow is of great significance also in the interpretation of natural phenomena.

Significance of shadows:

(1) Characteristic pattern caused by shadows of shrubs cast on an open field.

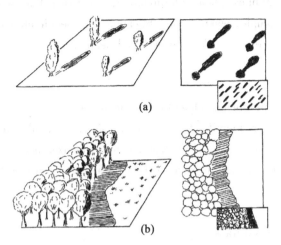

Figure 6.6　Shadow

(2) The shadow at the edge of a forest enhances the boundary between two different land types.

6.2.4　Pattern

Pattern refers to the arrangement of individual objects into distinctive recurring forms that facilitate the recognition on aerial imagery (Figure 6.7).

In land use studies, a number of patterns can be easily associated with human activities. This includes the checkerboard pattern of cultivation fields, the organized pattern of forest clear-cutting and urban area, and the linear patterns of highways and railways.

Figure 6.7　Pattern

6.2.5　Association

Association specifies the occurrence of certain objects of features, usually without the strict spatial arrangement implied by pattern.

In the context of military photo interpretation, the association of specific items has great significance, as, for example, when the identification of a specific class of equipment implies that other, more important, items are likely to be found nearby.

6.2.6 Shape

Shapes of features are obvious clues to their identities.

Typical examples for identification by their shapes include alluvial fans, anticlines, vegetation fields, and ocean waves. These features are recognized because of the experience of the interpreter in associating a certain form or shape with a specific surface feature.

6.2.7 Size

Size is important in two ways.

First, the relative size of an object or feature in relation to other objects in the image provides the interpreter with an intuitive notion of its scale and resolution, even though no measurements or calculations may have been made.

Second, absolute measurements can be valuable as interpretation.

6.2.8 Site

Site refers to topographic position. For example, sewage treatment facilities are positioned at lower topographic sites near streams or rivers to collect waste flowing through the system from high locations.

6.3 Image Interpretation Strategies

Generally, image interpretation includes the following steps:
- Field observations
- Direct recognition
- Interpretation by inference
- Probabilistic interpretation
- Deterministic interpretation

6.3.1 Field Observations

Field observations are required when the image and its relationship to ground conditions are so imperfectly understood that the interpreter is forced to go to the field to make an identification.

The analyst is unable to interpret the image from his or her knowledge and experience at hand, and must gather field observations to ascertain the relationship between the landscape and its appearance in the image.

6.3.2 Direct Recognition

Direct recognition is the application of an interpreter's experience, skill, and judgment to associate the image patterns with informational classes.

The process is essentially a qualitative, subjective analysis of the image using the elements of image interpretation as visual and logical clues.

6.3.3 Interpretation by Inference

Interpretation by inference is the use of a visible distribution to map one that cannot be observed in the image. The visible distribution acts as a surrogate, or substitute, for the mapped distribution.

For example, soils are defined by vertical profiles that cannot be directly observed by remotely sensed imagery. But soil distributions are sometimes very closely related to patterns of landforms and vegetation that are recorded in the image.

6.3.4 Probabilistic Interpretation

Probabilistic interpretation is an effort to narrow the range of possible interpretations by formally integrating non-image information into the classification process, often by means of quantitative classification algorithms.

6.3.5 Deterministic Interpretation

Deterministic interpretation is based on quantitatively expressed relationships that tie image characteristics to ground conditions.

In contrast with other methods, most information is derived from the image itself.

It is the rigorous and precise approach to image interpretation.

☞ Summary

- Concepts
- Task of image interpretation
- Elements of image interpretation
- Image interpretation strategies

☞ Key Points

- Image interpretation tasks
- Elements of image interpretation

☞ Review Questions

Do you agree or disagree the opinion—"visual interpretation is useless because of the emergence and widely application of digital interpretation"?

Chapter 7 Classification of Remotely Sensed Data

7.1 Introduction

Digital image classification is regarded as a fundamental process in remote sensing used to relate pixel values to land cover or sometimes land used classes that are present at the corresponding location on the Earth's surface. Usually each pixel is treated as an individual unit composed of values in several spectral bands. By comparing pixels to one another, and to pixels of known identity, it is possible to assemble groups of similar pixels into classes that are associated with the informational categories of interest to users of remotely sensed data. These classes form regions on a map or an image, so that after classification the digital image is presented as a mosaic of uniform parcels, each identified by a color or symbol (Figure 7.1). These classes are, in theory, homogeneous: pixels within classes are spectrally more similar to one another than they are to pixels in other classes. In practice, of course, each class will display some diversity, as each scene will exhibit some variability within classes.

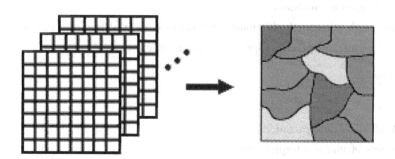

Figure 7.1 Numeric image and classified image

The classified image (right) is defined by examining the numeric image, then grouping together those pixels that have similar spectral values. Usually there are many more classes and at least three or four spectral bands

Image classification is an important part of the fields of remote sensing, image analysis, and pattern recognition. In some instances, the classification itself may be the object of the analysis. For example, remote sensing data is a primary source and used extensively for land use classification. Classification of land use produces a map-like image as the final product of the analysis. In other instances, the classification may be only an intermediate step in a more elaborate analysis, in which the classified data form one of several data layers in GIS. For example, in a study of water quality, an initial step may be to use image classification to identify wetlands and open water within a scene. Later steps may then focus upon more detailed study of these areas to identify influences upon water quality and to map variations in water quality. Image classification therefore forms an important tool for examination of digital images-sometimes to produce a final product, other times as one of several analytical procedures applied to derive information from an image.

The term classifier refers loosely to a computer program that implements a specific procedure for image classification. Over the years scientists have devised many classification strategies. There are several classification techniques/algorithms that are available, such as, supervised, unsupervised, decision tree or knowledge based, object oriented, artificial neural network, support vector machines and random forest classification techniques. However, no one ideal classification method exists and is unlikely that one could ever be developed. The analyst must select a classification method that will best accomplish a specific task. Therefore, it is essential that each analyst understand the alternative strategies for image classification so that he or she may be prepared to select the most appropriate classifier for the task at hand.

The simplest form of digital image classification is to consider each pixel individually, assigning it to a class based upon its several values measured in separate spectral bands (Figure 7.2). Sometimes such classifiers are referred to as spectral or point classifiers because they consider each pixel as a "point" observation (i.e., as values isolated from their neighbors). Although point classifiers offer the benefits of simplicity and economy, they are not capable of exploiting the information contained in relationships between each pixel and those that neighbor it. Human interpreters, for example, could derive little information using the point-by-point approach, because humans derive less information from the brightnesses of individual pixels than they do from the context and the patterns of brightnesses, of groups of pixels, and from the sizes, shapes, and arrangements of parcels of adjacent pixels.

As an alternative, more complex classification processes consider groups of pixels within their spatial setting within the image as a means of using the textural information so important for the human interpreter. These are spatial, or neighborhood, classifiers, which examine small areas within the image using both spectral and textural information to classify the image (Figure 7.3). Spatial classifiers are typically more difficult to program and much more expensive to use than point classifiers. In some situations spatial classifiers have demonstrated improved accuracy, but few have found their way into routine use for remote sensing image classification.

Another kind of distinction in image classification separates supervised classification from unsupervised classification. The supervised classification is the essential tool used for extracting quantitative information from remotely sensed image data. Using this method, the analyst has

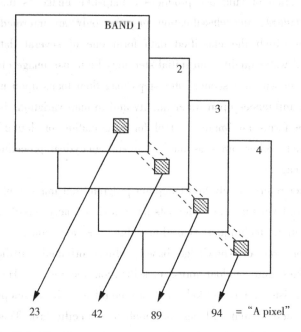

Figure 7.2 Point classifiers operate upon each pixel as a single set of spectral values considered in isolation from its neighbors

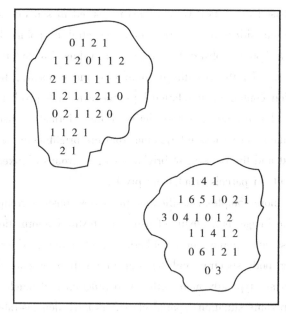

Figure 7.3 Image texture, the basis for neighborhood classifier. A neighborhood classifier considers values within a region of the image in defining class membership. Here two regions within the image differ with respect to average brightness and also with respect to "texture", the uniformity of pixels within a small neighborhood

available sufficient known pixels to generate representative parameters for each class of interest. This step is called training. One trained, the classifier is then used to attach labels to all the image pixels according to the trained parameters. Unsupervised classification, on the other hand, doesn't require human to have the foreknowledge of the classes, and mainly using some clustering algorithm to classify an image data. The distinction between supervised and unsupervised classification is useful, especially for students who are first learning about image classification. But the two strategies are not as clearly distinct as these definitions suggest, for some methods do not fit neatly into either category. These so-called hybrid classifiers share characteristics of both supervised and unsupervised methods.

7.2　Information Classes and Spectral Classes

Informational classes are the categories of interest to the users of the data. Informational classes are, for example, the different kinds of geological units, different kinds of forest, or the different kinds of land use that convey information to planners, managers, administrators, and scientists who use information derived from remotely sensed data. These classes are the information that we wish to derive from the data—they are the object of our analysis. Unfortunately, these classes are not directly recorded in remotely sensed images; we can derive them only indirectly, using the evidence contained in brightnesses recorded by each image. For example, the image cannot directly show geological units, but rather only the differences in topography, vegetation, soil color, shadow, and other factors that lead the analyst to conclude that certain geological conditions exist in specific areas.

Spectral classes are groups of pixels that are uniform with respect to the brightnesses in their several spectral channels. The analyst can observe spectral classes within remotely sensed data; if it is possible to define links between the spectral classes on the image and the informational classes that are of primary interest, then the image forms a valuable source of information. Thus, remote sensing classification proceeds by matching spectral categories to informational categories. If the match can be made with confidence, then the information is likely to be reliable. If spectral and informational categories do not correspond, then the image is unlikely to be a useful source for that particular form of information. Seldom can we expect to find exact one-to-one matches between informational and spectral classes. Any informational class includes spectral variations arising from natural variations within the class. For example, a region of the informational class "forest" is still "forest", even though it may display variations in age, species composition, density, and vigor, which all lead to differences in the spectral appearance of a single informational class. Furthermore, other factors, such as variations illumination and shadowing, may produce additional variations even within otherwise spectrally uniform classes.

Thus, informational classes are typically composed of numerous spectral subclasses, spectrally distinct groups of pixels that together may be assembled to form an informational class

(Figure 7.4). In digital classification, we must often treat spectral subclasses as distinct units during classification, but then display several spectral classes under a single symbol for the final image or map to be used by planners or administrators (who are, after all, interested only in the informational categories, not in the intermediate steps required to generate them).

7.3 Unsupervised Classification

Unsupervised classification procedures are characterized by the use of image properties to produce an initial definition of the cover types. This clustering methods attempt to find the underlying structure automatically by organizing the data into classes sharing similar, i.e., spectrally homogeneous, characteristics. The analyst "simply" needs to specify the number of clusters.

7.3.1 Advantages of Unsupervised Classifications

The advantages of unsupervised classification (relative to supervised classification) can be enumerated as follows:

The unsupervised method of classification is independent of external sources. The nature of knowledge required for unsupervised classification differs from that required for supervised classification. To conduct supervised classification, detailed knowledge of the area to be examined is required to select representative examples of each class to be mapped. To conduct unsupervised classification, no detailed prior knowledge is required, the classification of the image is done automatically and involves random sampling in unknown land-cover types.

The opportunity for human error is minimized. The operator may perhaps specify only the number of categories desired (or possibly, minimum and maximum limits on the number of categories), and sometimes constraints governing the distinctness and uniformity of groups. Almost no detailed decisions are required for unsupervised classification, so the analyst is presented with less opportunity for error. If the analyst has inaccurate preconceptions regarding the region, they will have little opportunity to influence the classification.

The unsupervised classification provides a useful method for organizing a large set of data so that the retrieval of information may be made more efficiency.

7.3.2 Disadvantages and Limitations

The disadvantages and limitations of unsupervised classification arise primarily from a reliance upon "natural" grouping and difficulties in matching these groups to the information categories that are of interest to the analyst.

(1) Unsupervised classification identifies spectrally homogeneous classes within the data that do not necessarily correspond to the informational categories that are of interest to the analyst. As a

result, the analyst is faced with the problem of matching spectral classes generated by the classification to the informational classes that are required by the ultimate user of the information. Seldom is there a simple one-to-one correspondence between the two sets of classes.

(2) The analyst has limited control over the menu of classes and their specific identities. If it is necessary to generate a specific menu of informational classes (e.g., to match to other classifications for other dates or adjacent regions), the use of unsupervised classification may be unsatisfactory.

(3) Accuracy may not be efficient because of no prior knowledge of the land cover types.

Unsupervised classification plays an especially important role when very little a prior information about the data is available. A primary objective of using clustering algorithms for multispectral remote sensing data is to obtain useful information for the selection of training regions in a subsequent supervised classification.

7.3.3 Distance Measures

Figure 7.4 shows two pixels, each with measurements in several spectral channels, plotted in multidimensional data space. For ease of illustration, only two bands are shown here, although the principles illustrated extend to as many bands as may be available. Unsupervised classification of an entire image must consider many thousands of pixels. But the classification process is always based upon the answer to the same question: "Do the two pixels belong to the same group?" For this example, the question is, "Should pixel C be grouped with A or with B?" This question can be answered by finding the distance between pairs of pixels. If the distance between A and C is greater than that between B and C, the B and C are said to belong to the same group and A may be defined as a member of a separate class.

There are thousands of pixels in a remotely sensed image; if they are considered individually as prospective members of groups, the distances to other pixels can always be used to define group membership. How can such distances be calculated? A number of methods for finding distances in multidimensional data space are available. One of the simplest is the Euclidean distance:

$$D_{AB} = \left[\sum_{i=1}^{n} (A_i - B_i)^2 \right]^{1/2} \qquad (7.1)$$

Where i is one of n spectral bands, A and B are pixels, and D_{AB} is the distance between the two pixels. The distance calculation is based upon the Pythagorean theorem:

$$c = \sqrt{a^2 + b^2} \qquad (7.2)$$

In this instance we are interested in distance c; a, b, and c are measured in units of the two spectral channels.

$$c = D_{AB} \qquad (7.3)$$

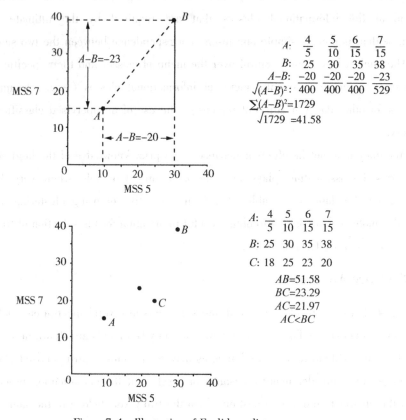

Figure 7.4 Illustration of Euclidean distance measure

To find D_{AB}, we need to find distance a and b. Distance a is found by subtracting the values of A and B in MSS channel 7 ($a = 38 - 15 = 23$). Distance b is found by finding the distance between A and B with respect to channel 6 ($b = 30 - 10 = 20$).

$$D_{AB} = c = \sqrt{20^2 + 23^2}$$
$$D_{AB} = \sqrt{400 + 529} = \sqrt{929}$$
$$D_{AB} = 30.47$$

This measure can be applied to as many dimensions (spectral channels) as might be available, by the addition of distances. For example,

The lower part of Figure 7.4 shows another worked example.

Thus the Euclidean distance between A and B is equal to 40.45 distance units. This value in itself has little significance, but in relation to other distances it forms a means of defining similarities between pixels. For example, if we find that distance $ab = 40.45$ and that distance $ac = 86.34$, then we know that pixel A is closer (i.e., more nearly similar) to B than it is to C, and that we should form a group from A and B rather than A and C.

	Landsat MSS band			
	1	2	3	4
Pixel A	34	28	22	6
Pixel B	26	16	52	29
Difference	8	12	−30	−23
(Difference)^2	64	144	900	529

Total of (differences)^2 = 1637

$\sqrt{\text{total}} = 40.5$

Unsupervised classification proceeds by making thousands of distance calculations as a means of determining similarities for the many pixels and groups within an image. Usually the analyst does not actually know any of these many distances that must be calculated for unsupervised classification, as the computer presents only the final classified image without the intermediate steps necessary to derive the classification. Nonetheless, distance measures are the heart of unsupervised classification.

But not all distance measures are based upon Euclidean distance. Another simple measure of determining distance is the L_1 distance, the sum of the absolute differences between values in individual bands (Swain and Davis, 1978). For the example given above, the L_1 distance is 7373 = (8+12+30+23). Other distance measures have been defined for unsupervised classification; many are rather complex methods of scaling distances to promote effective groupings of pixels.

7.3.4 Specific Methods for Unsupervised Classification

Typical known unsupervised classification algorithms have been developed for multi-spectral imagery, i.e., K means and ISODATA. These two algorithms are discussed here.

K Means Clustering

The K means clustering method is one of most common approaches used in image analysis applications. It requires an initial assignment of the available measurement vectors into a user-specified number of clusters, with arbitrarily specified initial cluster centers that are represented by the means of the pixel vectors assigned to them. This will generate a very crude set of clusters. The pixel vectors are then reassigned to the cluster with the closest mean, and the means are recomputed. The process is repeated as many times as necessary such that there is no further movement of pixels between clusters. The K means or iterative optimization algorithm is implemented in the following steps.

(1) Select a value for C, the number of clusters into which the pixels are to be grouped. This requires some feel beforehand as to the number of clusters that might naturally represent the image data set. Depending on the reason for using clustering some guidelines are available.

(2) Initialize cluster generation by selecting C points in spectral space to serve as candidate cluster centers, named $m_c(c = 1, 2, \cdots C)$. In principle the choice of the m_c at this stage is arbitrary with the exception that no two can be the same. To avoid anomalous cluster generation with unusual data sets it is generally best to space the initial cluster centers uniformly over the data. That can also aid convergence.

(3) Assign each pixel vectors x to the candidate cluster of the nearest mean using an appropriate distance metric in the spectral domain between the pixel and the cluster means. Euclidean distance is commonly used. That generates a cluster of pixel vectors about each candidate cluster mean.

(4) Compute a new set of cluster means from the groups formed in step 3; named those $n_c(c = 1, 2, \cdots C)$.

(5) If $n_c = m_c$ for all c then the procedure is complete. Otherwise the n_c are set to the current values of m_c and the procedure returns to step 3.

ISODATA Clustering

The ISODATA clustering algorithm builds on the k mean approach by introducing a number of checks on the clusters formed, either during or at the end of the iterative assignment process. Those checks relate to the number of pixels assigned to clusters and their shapes in the spectral domain.

At any suitable stage clusters can be examined to see whether:

(1) Any contain so few points as to be meaningless; for example, if the statistical distribution of pixels within clusters are important, as they might be when clustering is used as a pre-processing operation for maximum likelihood classification, sufficient pixels per cluster must be available to generate reliable mean and covariance estimates;

(2) Any are so close together that they represent an unnecessary of inappropriate division of the data, in which case they should be merged.

There is a guideline exists for (1). A cluster cannot reliably be modeled by a multivariate normal distribution unless it contains about 10N numbers, where N is the number of spectral components. Decisions in (2) about when to merge adjacent clusters can be made by assessing how similar they are spectrally. Similarity can be assessed simply by the distance between them in the spectral domain.

Another test sometimes incorporated in the ISODATA algorithm concerns the shapes of clusters in spectral space. Clusters that are elongated can be split in two, if required. Such a decision can be made on the basis of pre-specifying a standard deviation in each spectral band beyond which a cluster should be halved.

7.3.5 Sequence for Unsupervised Classification

A typical sequence might begin with the analyst specifying minimum and maximum numbers of categories to be generated by the classification algorithm. These values might be based upon the analyst's requirement that the final classification display a certain number of classes. The classification starts with a set of arbitrarily selected pixels as cluster centers; often these are selected at random to ensure that the analyst cannot influence the classification and that the selected pixels are representative of values found throughout the scene. The classification algorithm then finds distances (as described above) between pixels and forms initial estimates of cluster centers as permitted by constraints specified by the analyst. The class can be represented by a single point, known as the "class centroid," which can be thought of as the center of the cluster of pixels for a given class, even though many classification procedures do not always define it as the exact center of the group. At this point, classes consist only of the arbitrarily selected pixels chosen as initial estimates of class centroids. In the next step, all the remaining pixels in the scene are assigned to the nearest class centroid. The entire scene has now been classified, but this classification forms only an estimate of the final result, as the classes formed by this initial attempt are unlikely to be the optimal set of classes and may not meet the constraints specified by the analyst.

To begin the next step, the algorithm finds new centroids for each class, as the addition of new pixels to the classification means that the initial centroids are no longer accurate. Then the entire scene is classified again, with each pixel assigned to the nearest centroid. And again new centroids are calculated; if the new centroids differ from those found in the preceding step, then the process repeats until there is no significant change detected in locations of class centroids and the classes meet all constraints required by the operator.

Throughout the process the analyst generally has no interaction with the classification, so it operates as an "objective" classification within the constraints provided by the analyst. Also, the unsupervised approach identifies the "natural" structure of the image in the sense that it finds uniform groupings of pixels that form distinct classes without the influence of preconceptions regarding their identities or distributions. The entire process, however, cannot be considered to be "objective," as the analyst has made decisions regarding the data to be examined, the algorithm to be used, the number of classes to be found, and (possibly) the uniformity and distinctness of classes. Each of these decisions influences the character and the accuracy of the final product, so it cannot be regarded as a result isolated from the context in which it was made.

Many different procedures for unsupervised classification are available; despite their diversity, most are based upon the general strategy just described. Although some refinements are possible to improve computational speed and efficiency, this approach is in essence a kind of wearing down of the classification problem by repetitive application assignment and reassignment of pixels to groups. Key components to any unsupervised classification algorithm are effective methods of measuring distances in data space, identifying class centroids, and testing the distinctness of classes. There are many different strategies for accomplishing each of these tasks; an enumeration of even the most widely used methods is outside the scope of this text, but some are described in relative articles.

7.4 Supervised Classification

Supervised classification can be defined informally as the process of using samples of known identity (i.e., pixels already assigned to informational classes) to classify pixels of unknown identity (i.e., to assign unclassified pixels to one of several informational classes). Samples of known identity are those pixels located within training areas, or training fields. The analyst defines training areas by identifying regions on the image that can be clearly matched to areas of known identity on the image. Such areas should typify spectral properties of the categories they represent, and, of course, must be homogeneous with respect to the informational category to be classified. That is, training areas should not include unusual regions, nor should they straddle boundaries between categories. Size, shape, and position must favor the convenient identification both on the image and on the ground. Pixels located within these areas form the training samples used to guide the classification algorithm to assign specific spectral values to appropriate informational classes. Clearly, the selection of these training data is a key step in supervised classification.

7.4.1 Advantages of Supervised Classifications

The advantages of supervised classification, relative to unsupervised classification, can be enumerated as follows:

First, the analyst has control of a selected menu of informational categories tailored to a specific purpose and geographic region. This quality may be vitally important if it becomes necessary to generate a classification for the specific purpose of comparison with another classification of the same area at a different date or if the classification must be compatible with those of neighboring regions. Under such circumstances, the unpredictable (i.e., with respect to number, identity, size, and pattern) qualities of categories generated by unsupervised classification may be inconvenient or unsuitable.

Second, supervised classification is tied to specific areas of known identity, determined through the process of selecting training areas.

Third, the analyst using supervised classification is not faced with the problem of matching spectral categories on the final map with the informational categories of interest (this task has, in effect, been addressed during the process of selecting training data).

Finally, the operator may be able to detect serious errors in classification by examining training data to determine if they have been correctly classified by the classification procedure-inaccurate classification of training data indicates serious problems in the classification or selection of training data, although the correct classification of training data does not always indicate the correct classification of other data.

7.4.2 Disadvantages of Superised Classification

The disadvantages of supervised classification are numerous.

First, the analyst, in effect, imposes a classification structure upon the data (recall that unsupervised classification searches for "natural" classes). These operator-defined classes may not match the natural classes that exist within the data, and therefore may not be distinct or well defined in multidimensional data space.

Second, training data are often defined primarily with reference to informational categories and only secondarily with reference to spectral properties. A training area that is "100% forest" may be accurate with respect to the "forest" designation, but may still be very diverse with respect to density, age, shadowing, and the like, and therefore form a poor training area.

Third, training data selected by the analyst may not be representative of conditions encountered throughout the image. This may be true despite the best efforts of the analyst, especially if the area to be classified is large, complex, or inaccessible.

Fourth, conscientious selection of training data can be a time-consuming, expensive, and tedious undertaking, even if ample resources are at hand. The analyst may have problems in matching prospective training areas as defined on maps and aerial photographs to the image to be classified.

Finally, supervised classification may not be able to acknowledge and represent special or unique categories not represented in the training data, possibly because they are not known to the analyst or because they occupy very small areas on the image.

7.4.3 Training Data

Training fields are areas of known identity delineated on the digital image, usually by specifying the corner points of a square or rectangular area using line and column numbers within the coordinate system of the digital image. The analyst must, of course, know the correct class for each area. Usually the analyst begins by assembling and studying maps and aerial photographs of the area to be classified and by investigating selected sites in the field. (Here we assume that the analyst has some field experience with the specific area to be studied, is familiar with the particular problem the study is to address, and has conducted the necessary field observations prior to initiating the actual selection of training data.) Specific training areas are identified for each informational category, following the guidelines outlined below. The objective is to identify a set of pixels that accurately represent spectral variation present within each informational region (Figure 7.5).

1. Key Characteristics of Training Areas

The choice of training data is arguably the most difficult and critical part of the supervised classification process. The standard procedure is to select areas within a scene which are representative of each class of interest.

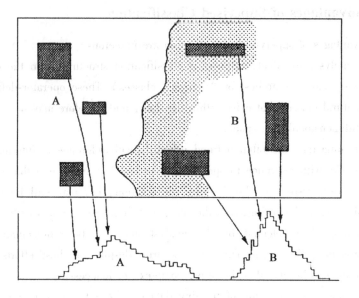

Figure 7.5　Training fields and training data

Training fields, each composed of many pixels, sample the spectral features of informational categories. Here the shaded figures represent training fields, each positioned carefully to estimate the spectral properties of each class, as represented by the histograms. This information provides the basis for classification of the remaining pixels not within the training areas

Number of Pixels

An important concern is the overall number of pixels selected for each category.

Chen et al. (2002) indicated that training numbers of 25 pixels for each class were apparently too small to extract the representative statistics. The training sizes with 50~100 pixels might be better for coarse resolution images. Joyce (1978) specifically refered to Landsat MMS data. In terms of numbers of pixels, he is recommending from 10 to about 40 pixels for each training field. For TM and SPOT data, of course, his recommendations would specify different numbers of pixels, as the resolutions of these sensors differ from that of the MSS. Also, since the optimum sizes of training fields vary according to the heterogeneity of each landscape and each class, each analyst should develop his or her own guidelines based on experience acquired in specific circumstances.

Size

As a general guideline, for an N dimensional spectral space, the minimum number of independent training pixels required is (N+1). Because of the difficult in assuring independence of the pixels, usually many more than this minimum number are selected. A practical minimum of 10N training pixels per spectral class is recommended, which as many as 100N per class if possible.

Shape

Shapes of training areas are not important, provided that shape does not prohibit accurate delineation and positioning of correct outlines of regions on digital images. Usually it is easiest to define square or rectangular areas; such shapes minimize the number of vertices that must be specified, usually the most bothersome task for the analyst.

Location

Location is important, as each informational category should be represented by several training areas positioned throughout the image. Training areas must be positioned in locations that favor accurate and convenient transfer of their outlines from maps and aerial photographs to the digital image. As the training data intended to represent variation within the image, they must not be clustered in favored regions of the image that may not typify conditions encountered throughout the image as a whole. It is desirable for the analyst to use direct field observations in the selection of training data, but the requirement of training data often conflicts with practical constraints, as it may not be practical to visit remote or inaccessible sites that may seem to from good areas for training data. Often aerial observation, or use of good maps and aerial photographs, can provide the basis for accurate delineation of training fields that inspected in the field. Although such practices are often sound, it is important to avoid development of a cavalier approach to selection of training data that depends completely upon indirect evidence in situations when direct observations are feasible.

Number

The optimum number of training areas depends upon the number of categories to be mapped, their diversity, and the resources that can be devoted to delineating training areas. Ideally, each informational category or each spectral subclass should be represented by a number (5 to 10 at a minimum) of training areas to ensure that the spectral properties of each category are represented. Chen et al (2002) indicated that 25 blocks for each class achieved the best classification accuracy for both coarser and finer resolution levels. Because informational classes are often spectrally diverse, it may be necessary to use several sets of training data for each informational category, due to the presence of spectral subclasses. Selection of multiple training areas is also desirable because later in the classification process it may be necessary to discard some training areas if they are discovered to be unsuitable. Experience indicates that it is usually better to define many small training areas than to use only a few large areas.

Placement

Placement of training areas may be important. Training areas should be placed within the image in a manner that permits convenient and accurate location with respect to distinctive features, such as water bodies, or boundaries between distinctive features on the image. They should be distributed throughout the image so that they provide a basis for representation of the diversity present within the scene. Boundaries of training fields should be placed well away from the edges of contrasting parcels so that they do not encompass edge pixels.

Uniformity

Perhaps the most important property of a good training area is its uniformity, or homogeneity. Data within each training area should exhibit a unimodal frequency distribution for each spectral band to be used (Figure 7.6). Prospective training areas that exhibit bimodal histograms should be discarded if their boundaries cannot be adjusted to yield more uniformity. Training data provide values that estimate the means, variances, and covariances of spectral data measured in several spectral channels. For each class to be mapped, these estimates approximate the mean values for each band, the variability of each band, and the interrelationships between bands. As an ideal, these values should represent the conditions present within each class within the scene, and thereby form the basis for classification of the vast majority of pixels within each scene that do not belong to training areas. In practice, of course, scenes vary greatly in complexity, and individual analysts differ in their knowledge of a region and in their ability to define training areas that accurately represent the spectral properties of informational classes. Moreover, some informational classes are not spectrally uniform and cannot be conveniently represented by a single set of training data.

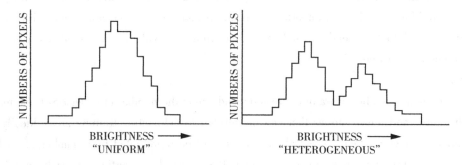

Figure 7.6 Uniforms and heterogeneous training data

On the left, the histogram of the training data has a single peak, indicating a degree of spectral homogeneity. Data from such training fields forms a suitable basis for image classification. On the right, a second set of training data displays a bimodal histogram that reveals that this area encompasses two, rather than one, spectral classes. This training area is not satisfactory for image classification and must be discarded or redefined

2. Significance of Training Data

Scholz, et al. (1979) and Hixson, et al. (1980) discovered that selection of training data may be as important as or even more important than choice of a classification algorithm in determining classification accuracies of agricultural areas in the central United States. They concluded that differences in the selection of training data were more important influences upon accuracy than were differences among some five different classification procedures. The results of their studies show little difference in the classification accuracies achieved by the five classification algorithms that were considered if the same training statistics were used. However, in one part of the study, a classification algorithm given two alternative training methods for the same data produced significantly different results. This finding suggests that the choice of training method, at least in some instances, is as important as the choice of classifier. Scholz, et al. (1979, p. 4) concluded that the most important aspect of training fields is that all cover types in the scene must be adequately represented by a sufficient number of samples in each spectral subclass.

A study by Campbell (1981) examined the character of the training data as it influences the accuracy of the classification. His examples showed that adjacent pixels within training fields tended to have similar values; as a result, the samples that compose each training field may not be independent samples of the properties within a given category. Training samples collected in contiguous blocks may tend to underestimate the variability within each class and to overestimate the distinctness of categories. His examples also show that the degree of similarity varies between land-cover categories, from band to band, and from date to date. If training samples are selected randomly within classes, rather than as blocks of contiguous pixels, effects of high similarity are minimized and classification accuracies improve. Also, his results suggest that it is probably better to use a rather large number of small training fields rather than a few large areas.

3. Idealized Sequence for Selecting, Training Data

Specific circumstances for conducting supervised classification vary greatly, so it is not possible to discuss in detail the procedures to follow in selecting, training data, which will be determined in part by the equipment and software available at a given facility. However, it is possible to outline an idealized sequence as a suggestion of the key steps in the selection and evaluation of training data.

(1) Assemble information, including maps and aerial photographs of the region to be mapped.

(2) Conduct field studies, to acquire first-hand information regarding the area to be studied. The amount of effort devoted to field studies varies depending upon the analyst's familiarity with the region to be studied. If the analyst is intimately familiar with the region and has access to up-to-date maps and photographs, additional field observations may not be necessary. However, in the vast majority of situations at least some field work will be required.

(3) Carefully plan collection of field observations, and choose a route designed to observe all regions of the study region. Maps and images should be taken into the field in a form that permits the analyst to annotate them as he or she makes observations in the field (e.g., images may be prepared with overlays; photocopies of maps can be marked with colored pens). Although time and access may be limited, it is important to observe all classes of terrain encountered within the study area, as well as all regions. Observations should not be limited to a few easily observed segments of the area. The analyst should keep good notes, carefully keyed to the annotations on the image. He or she may find it useful to take photographs as a permanent record of conditions observed. In very remote areas, aerial observation may be the only practical means of observing

the study region. Ideally, field observations should be timed to coincide with image acquisition; when this is not possible, they should at least be made during the season images were acquired.

(4) Conduct a preliminary examination of the digital scene. Determine landmarks that may be useful in positioning training fields. Assess image quality. Examine frequency histograms of the data, and determine if preprocessing is necessary.

(5) Identify prospective training areas, using guidelines proposed by Joyce (1978) and outlined here. Sizes of prospective areas must be assessed in the light of scale differences between maps or photographs and the digital image. Locations of training areas must be defined with respect to features easily recognizable on the image and on the maps and photographs used as collateral information.

(6) Display the digital image, then locate and delineate training areas on the digital image. Be sure to place training area boundaries well inside parcel boundaries to avoid including mixed pixels within training areas. At the completion of this step, all training areas should be identified with respect to row and column coordinates within the image.

(7) For each training area, display and inspect frequency histograms of all spectral bands to be used in the classification. If possible, estimate the means, variance, dispersion, covariance, and so on, to assess the usefulness of training data.

(8) Modify boundaries of the training areas to eliminate bimodal frequency distributions, or, if necessary, discard those areas that are not suitable. If necessary, return to Step 1 to define new areas to replace those that have been eliminated.

(9) Incorporate training data information into a form suitable for use in the classification procedure and proceed with the classification process as described in subsequent sections of this chapter.

7.4.4 Specific Methods for Supervised Classification

A variety of different methods have been devised to implement the basic strategy of supervised classification. Use all information derived from the training data as a means of classifying those pixels not assigned to training fields. Various classifiers of supervised classification have been developed for assigning an unknown pixel to one of the information classes. Among the most frequently used are the parallelepiped, minimum distance, and maximum likelihood classifiers.

1. Parallelepiped Classification

The parallelepiped classifier first calculates the range of DN values in each category of training data sets. An unknown pixel is assigned to a class in which it lies. If a pixel is located outside all class ranges, it will be classified as "unknown". Figure 7.7 is the schematic concept for the parallel piped classifier.

An example can be formed from data shown in Table 7.1. Here Landsat MSS bands 5 and 7 are taken from a larger data set to provide a concise, easily illustrated example. In practice, four or more bands can be used. The ranges of values with respect to band 5 can be plotted on the horizontal axis in Figure 7.8. The extremes of values in band 7 training data are plotted along the vertical axis, and then projected to intersect with the ranges from band 5. The polygons thus defined (Figure 7.8) represent regions in data space that are assigned to categories in classification. As pixels of unknown identity are considered for classification, those that fall in these regions are assigned to the category associated with each polygon, as derived from training data. The procedure can be extended to as many bands, or as many categories as necessary. In addition, the decision boundaries can be defined by the standard deviations of the values within the training areas rather than their ranges. This kind of strategy is useful because fewer pixels will be placed in an "unclassified" category (a special problem for parallelepiped classification), but it also increases the opportunity for classes to overlap in spectral data space.

Table 7.1 **Data for example shown in Figure 7.8**

	Group A				Group B			
	Band				Band			
	4	5	6	7	4	5	6	7
	34	28	22	3	28	18	59	35
	36	35	24	6	28	21	57	34
	36	28	22	6	28	21	57	30
	36	31	23	5	28	14	59	35
	36	34	25	7	30	18	62	28
	36	31	21	6	30	18	62	38
	35	30	18	6	28	16	62	36
	36	33	24	2	30	22	59	37
	36	36	27	10	27	16	56	34
High	34	28	18	10	27	14	56	28
Low	36	36	27	3	30	22	62	38

Note. These data have been selected from a larger data set to illustrate parallelepiped classification.

Although this procedure for classification has the advantages of accuracy, directness, and simplicity, some of its disadvantages are obvious. The main difficulty arises when two or more category ranges overlap. Unknown pixels that lie in the overlapped areas will be assigned arbitrarily to one of the overlapping classes or be labeled as "unknown". This may be caused by the following reasons. Training data may underestimate actual ranges of classification and leave large areas in data space and on the image unassigned to informational categories. Also, the

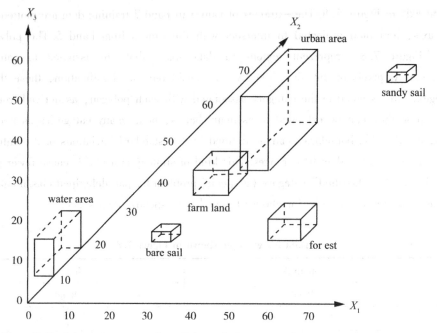

Figure 7.7 Schematic concept of parallel piped classifier in three dimensional feature space

regions as defined in data space are not uniformly occupied by pixels in each category; those pixels near the edges of class boundaries may belong to other classes. Also, if training data do not encompass the complete range of values encountered in the image (as is frequently the case) large areas of the image remain unclassified, or the basic procedure described here must be modified to assign these pixels to logical classes.

2. Minimum Distance Classification

The minimum distance classifier computes the mean vectors of the spectral data for each class based on the training data set. The spectral data from training fields can be plotted in multidimensional data space in the same manner illustrated previously for unsupervised classification. Values in several bands determine the positions of each pixel within the clusters that are formed by training data for each category (Figure 7.9). These clusters may appear to be the same as those defined earlier for unsupervised classification. However, in unsupervised classification, these clusters of pixels were defined according to the "natural" structure of the data. Now, for supervised classification, these groups are formed by values of pixels within the training fields defined by the analyst.

Each cluster can be represented by its centroid, often defined as its mean value. As unassigned pixels are considered for assignment to one of the several classes, the multidimensional distance to each cluster centroid is calculated and the pixel is then assigned to the closest cluster. Thus the classification proceeds by always using the "minimum distance" from a given

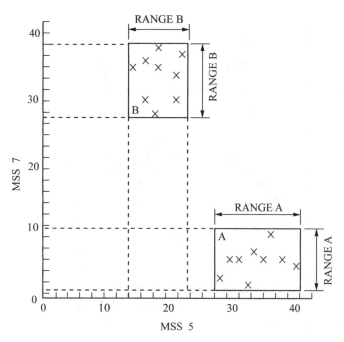

Figure 7.8 Parallelepiped classification Ranges of values within training data (Table 7.1) define decision boundaries

Here only two bands are shown, but the principle extends to several spectral bands. Other pixels, not from the training fields, are classified as a given category if their positions fall within the polygons defined by the training data

pixel to a cluster centroid defined by the training data as the spectral manifestation of an informational class.

Minimum distance classifiers are direct in concept and in implementation, but are not widely used in remote sensing work. This classifier is not sensitive to different degrees of variance in the spectral response data. Therefore, the classifier is not applicable where spectral classes are close in the feature space and are of high variance. It is possible to devise more sophisticated versions of the basic approach just outlined by using different distance measures and different methods of defining cluster centroids.

3. Maximum Likelihood Classification

In nature the classes that we classify exhibit natural variation in their spectral patterns. Further variability is added by the effects of haze, topographic shadowing, system noise, and the effects of mixed pixels. As a result, remote sensing images seldom record spectrally pure classes; more typically, they display a range of brightnesses in each band. The classification strategies considered thus far do not consider variation that may be present within spectral categories, and do not address problems that arise when frequency distributions of spectral values from separate

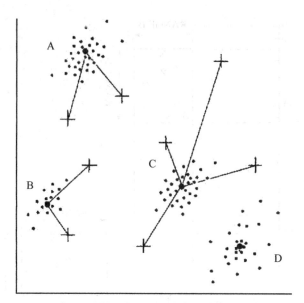

Figure 7. 9 Minimum distance classifier

Here small dots represent pixels from training fields and crosses indicate the large unclassified pixels from elsewhere in the image; each of these pixels is assigned to the class with the closest centroid, as measured using the distance measures discussed in the text

categories overlap. For example, for application of a parallelepiped classifier, the overlap of classes is a serious problem because spectral data space cannot then be neatly separated into discrete units for classification. This kind of situation arises frequently because often our attention is focused upon classifying those pixels that tend to be spectrally similar rather than those that are distinct enough to be easily and accurately classified by other classifiers.

As a result, the situation depicted in Figure 7. 10 is common. Assume that we examine a digital image representing a region composed of three-fourths forested land and one-fourth cropland. The two classes "Forest" and "Cropland" are distinct with respect to average brightness, but extreme values (very bright forest pixels or very dark crop pixels) are similar in the region where the two frequency distributions overlap. (For clarity, Figure 7. 10 shows data for only a single spectral band, although, the principle extends to values observed in several bands and to more than the two classes shown here.) Brightness value "45" falls into the region of overlap; where we cannot make the kinds of decision rules clear assignment to either "Forest" or to "Cropland". Using the kinds of decision rules mentioned above, we cannot decide which group should receive these pixels unless we place the decision boundary arbitrarily.

In this situation, an effective classification would consider the relative likelihoods of "45 as a member of forest" and "45 as a member of Cropland". We could then choose the class that would maximize the probability of a correct classification, given the information in the training data. This kind of strategy is known as maximum likelihood classification—it uses the training data as a means of estimating means and variances of the classes, which are then used to estimate the probabilities. Maximum likelihood classification considers not only the mean, or average, values in assigning classification, but also the variability of brightness values in each class.

The maximum likelihood classifier assumes that the training data statistics for each class in each band are normally distributed, that is, the Gaussian distribution. Under this assumption, the distribution of spectral response pattern of a class can be measured by the mean vector and the covariance matrix. To classify an unknown pixel, the maximum likelihood classifier computes the probability value of the pixel belonging to each class, and assigns it to the class that has the largest (or maximum) value. A pixel will be labeled "unknown" if the probability values are all below the threshold values set by the analyst. It requires intensive calculations, so it has the disadvantage of requiring more computer resources than do most of the simpler techniques mentioned above. Also, it is sensitive to variations in the quality of the training data-even more so than most other supervised techniques. Computation of the estimated probabilities is based on the assumption that both training data and the classes themselves display multivariate normal (Gaussian) frequency distributions. (This is one reason that training data should exhibit unimodal distributions, as discussed above.) Data from remotely sensed images often do not strictly adhere to this rule, although the departures are often small enough that the usefulness of the procedure is preserved. Nonetheless, training data that are not carefully selected may introduce error.

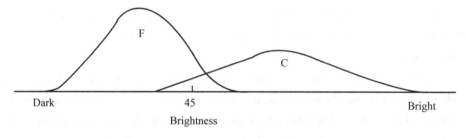

Figure 7.10 Maximum likelihood classification

These frequency distributions represent pixels from two training fields; the zone of overlap depicts pixel values common to both categories. The relations of the values within the overlap region to the overall frequency distribution for each class forms the basis for assigning pixels to classes

7.4.5 Idealized Sequence for Conducting Supervised Classification

The practice of supervised classification is considerably more complex than unsupervised classification. The analyst must evaluate the situation at each of several steps, and then return to an earlier point if it appears that refinements or corrections are necessary to ensure an accurate result.

(1) Prepare the menu of categories to be mapped. Ideally, these categories correspond to

those of interest to the final users of the maps and data. But these requirements may not be clearly defined, or the user may require the assistance of the analyst to prepare a suitable plan for image classification.

(2) Select and define training data, as sketched above. Select known, representative pixels for each of the classes. Those pixels are said to form training data. Training sets for each class can be established using site visits, maps, air photographs or even photo interpretation of image products formed from the data. Sometimes the training pixels for a given class will lie in a common region enclosed by a border. That region is called a training field.

This step requires careful comparison of the training fields as marked on maps and photographs with the digital image as displayed on a video screen. It is for this reason that training fields must be easily located with respect to distinctive features visible both on maps and on the digital image. The image-processing system then uses the image coordinates of the edges of the polygons to determine pixel values within the training field to be used in the classification process. After locating all training fields, the analyst can then display frequency distributions for training data as a means of evaluating the uniformity of each training field, and can use the training data to estimate the parameters of the particular classifier algorithm to be employed. The set of parameters is sometimes called the signature of that class.

(3) Modify categories and training fields as necessary to define homogeneous training data. As the analyst examines the training data, he or she may need to delete some categories from the menu, combine others, or to subdivide still others into spectral subclasses. If the training data for the modified categories meet the requirements of size and homogeneity, they can be used to enter the next step of the classification. Otherwise, the procedure (at least for certain categories) must start again at Step 1.

(4) Conduct classification. Each image analysis system will require a different series of commands to conduct a classification; but in essence the analyst must provide the program with access to the training data (often written to a specific computer file) and must identify the image that is to be examined. The results can be displayed on a video display terminal, and the thematic maps and tables which summarise class memberships of all pixles in the image should be produced, from which the areas of the classes can be measured.

(5) Evaluate classification performance. Finally, the analyst must conduct an evaluation using the procedures discussed in the next chapter.

This sequence outlines the basic steps required for supervised classification; the details may vary from individual to individual and with different image-processing systems, but the basic principles and concerns remain the same: accurate definition of homogenous training fields.

7.5 Textural Classifiers

Image classification presents special problems for digital analysis because of the diverse spectral characteristics of the landscape. Idealized landscape regions consist of spectrally homogeneous patches on the Earth's surface. Accurate mapping of such regions could be accomplished by the relatively straightforward process of matching spectral categories to the spectral "signatures" of informational categories, as described above.

No account is taken of how their neighbors might be labeled in the classifiers treated above. In any real image adjacent pixels are related or correlated because imaging sensors acquire significant portions of energy from adjacent pixels and because ground cover types generally occur over a region that is large compared with the size of a pixel. Low-density residential land, for example, as viewed from above in detail, is composed largely of tree crowns, rooftops, lawns, paved streets, driveways, and parking lots. We are interested in the classification of the composite of these many features rather than in making an inventory of the many components that in themselves may be of little interest. Thus, in the ideal, the classification should focus upon the overall pattern of variation that characterizes each category, rather than upon the average brightness, which may not reveal much about the essential differences between categories. Although human interpreters can intuitively recognize such complex patterns, many digital classification algorithms encounter serious problems in accurate classification of such scenes because they are designed to classify each separate spectral region as a separate informational category.

Classification methods that take into account the labeling of neighbors when seeking to determine the most appropriate class for a pixel are said to be context sensitive and are called context classifier. There are several approaches to perform the context classification, the most common of which are context classification by image pre-processing , e. g. , a median filter and post classification filtering. For example, the standard deviation of brightness values within a neighborhood of specified size, systematically moved over the entire image, may provide a rough measure of the spectral variability over short distances as a measure of image texture. Such a measure may permit the analyst to classify composite categories such as the one mentioned above.

Usually more sophisticated measures of texture are required to produce satisfactory results. For example, other textural measures, examine relationships between brightness values at varying distances and directions from a central pixel, which is systematically moved over the image (Haralick et al. , 1973; Maurer, 1974; Haralick, 1979).

Jensen (1979) found that the use of textural measures improved his classification of Level II and III suburban and transitional land using band 5 of the Landsat MSS. Improvements were confined to certain categories, and were perhaps of marginal significance when considered in the context of increased costs.

Textural measures seem to work best when relatively large neighborhoods are defined (perhaps as large as 64×64 pixels). Such large neighborhoods may cause problems when they straddle the boundaries between categories. In addition, such large neighborhoods may decrease the effective spatial resolution of the final map, as they must form its smallest spatial elements.

7.6 Fuzzy Clustering

Fuzzy clustering addresses a problem implicit in much of the preceding material: pixels must be assigned to a single discrete class. Although such classification attempts to maximize correct classifications, the logical framework allows only for direct one-to-one matches between pixels and classes. We know, however, that many processes contribute to prevent clear matches between pixels and classes. Therefore, the focus upon finding discrete matches between the pixels and informational classes ensures that many pixels will be incorrectly or illogically labeled. Fuzzy logic attempts to address this problem by applying a different classification logic.

Fuzzy logic (Kosko and Isaka, 1993) has applications in many fields, but has special significance for remote sensing. Fuzzy logic permits partial membership, a property that is especially significant in the field of remote sensing, as partial membership translates closely to the problem of mixed pixels. So whereas traditional classifiers must label pixels as either "Forest" or "Water", for example, a fuzzy classifier is permitted to assign a pixel a membership grade of 0.3 for "Water" and 0.7 for "Forest", in recognition that the pixel may not be properly assigned to a single class. Membership grades typically vary from 0 (nonmembership) to 1.0 (full membership), with intermediate values signifying partial membership in one or more other classes (Table 7.2).

Table 7.2   **Partial membership in fuzzy classes**

Class	\multicolumn{7}{c}{Pixel}
	A	B	C	D	E	F	G
Water	0.00	0.00	0.00	0.00	0.00	0.00	0.00
Urban	0.00	0.01	0.00	0.00	0.00	0.00	0.85
Transportation	0.00	0.35	0.00	0.00	0.99	0.79	0.14
Forest	0.07	0.00	0.78	0.98	0.00	0.00	0.00
Pasture	0.00	0.33	0.21	0.02	0.00	0.05	0.00
Cropland	0.92	0.30	0.00	0.00	0.00	0.15	0.00

A fuzzy classifier assigns membership to pixels based upon a membership function (Figure 7.11). Membership functions for classes are determined either by general relationships or by definitional rules describing the relationships between data and classes. Or, as is more likely in the instance of remote sensing classification, membership functions are derived from experimental (i.e., training) data for each specific scene to be examined. In the instance of remote sensing data, a membership function describes the relationship between class membership and brightness in several spectral bands (Figure 7.11).

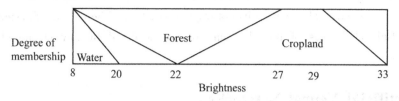

Figure 7.11 Membership functions for fuzzy clustering

Figure 7.11 provides contrived examples showing several pixels and their membership grades. (The actual output from a fuzzy classification is likely to form an image that shows varied levels of membership for specific classes.) Membership grades can be hardened (Table 7.3) by setting the highest class membership to 1.0 and all others to 0.0. Hardened classes are equivalent to traditional classifications: Each pixel is labeled with a single label and the output is a single image labeled with the identifier of the hardened class. Programs designed for remote sensing applications (Bezdek et al., 1984) provide the ability to adjust the degree of fuzziness and thereby adjust the structure of classes and the degree of continuity in the classification pattern.

Table 7.3　　　　"Hardened" classes for example shown in Table 7.2

Class	Pixel						
	A	B	C	D	E	F	G
Water	0.00	0.00	0.00	0.00	0.00	0.00	0.00
Urban	0.00	0.00	0.00	0.00	0.00	0.00	1.00
Transportation	0.00	1.00	0.00	0.00	1.00	1.00	0.00
Forest	0.07	0.00	1.00	1.00	0.00	0.00	0.00
Pasture	0.00	0.00	0.00	0.00	0.00	0.00	0.00
Cropland	1.00	0.00	0.00	0.00	0.00	0.00	0.00

Fuzzy clustering has been judged to improve results, at least marginally, with respect to traditional classifiers, although the evaluation is difficult because the usual evaluation methods require the discrete logic of traditional classifiers. Thus, the improvements noted for hardened classifications are probably conservative as they do not reveal the full power of fuzzy logic.

This example illustrates membership functions for the simple instance of three classes considered for a single band, although the method is typically applied to multiple bands. The horizontal axis represents pixel brightness; the vertical axis represents the degree of membership, from low near the bottom to high at the top. The class "Water" consists of pixels darker than brightness 20, although only pixels darker than 8 are likely to be completely occupied by open water. The class "Cropland" can include pixels as dark as 22 and as bright as 33, although pure "Cropland" pixels are found only in the range 27 to 29. A pixel of brightness 28, for example, can only be "Cropland," although a pixel of brightness 24 could be partially forested, partially in agriculture. Unlabeled areas of this diagram are not occupied by any of the classes in this classification.

7.7 Artificial Neural Networks

Artificial neural networks (ANNs) are computer programs that are designed to simulate human learning processes through the establishment and reinforcement of linkages between input data and output data. It is these linkages, or pathways, that form the analogy with the human learning process, in that repeated the association between input and output in the training process reinforce linkages, or pathways, that can then be employed to link input and output, in the absence of training data.

ANN makes no strong assumptions about the form of the probability distributions and can be adjusted flexibly to the complexity of the system that they are being used to model. They therefore provide an attractive compromise.

ANNs are often represented as composed of three elements. An input layer consists of the source data, which in the context of remote sensing are the multispectral observations, perhaps in several bands and from several dates. ANNs are designed to work with large volumes of data, including many bands and dates of multispectral observations, together with related ancillary data.

The output layers consist of the classes required by the analyst. There are few restrictions on the nature of the output layer, although the process will be more reliable when the number of output layers is small, or modest, with respect to the number of input channels. Included are training data in which the association between output layers and input data is clearly established. During the training phase, an ANN establishes an association between input and output data by establishment of weights within one or more hidden layers (Figure 7.12). In the context of remote sensing, repeated associations between classes and digital values, as expressed in the training data, strengthen weights within hidden layers that permit the ANN to assign correct labels when given spectral values in the absence of training data.

Further, ANNs can also be trained by back propagation (BP). If the establishment of the usual training data for conventional image classification can be thought of as "forward propagation," then BP can be thought of as a retrospective elimination of the links between input and output data in which differences between expected and actual results can be used to adjust the weights. This process establishes transfer functions, quantitative relationships between input and output layers that assign weights to emphasize effective links between input and output layers. For example, such weights might acknowledge that some band combinations may be very effective in defining certain classes and others effective for other classes. In BP, hidden layers note errors in matching data to classes and adjust the weights to minimize errors.

ANNs are designed using less severe statistical assumptions than many of the usual classifiers (e.g., maximum likelihood), although in practice successful application requires careful application. ANNs have been found to be accurate in the classification of remotely sensed data, although improvements inaccuracies have generally been small or modest. ANNs require considerable effort to train and are intensive in their use of computer time.

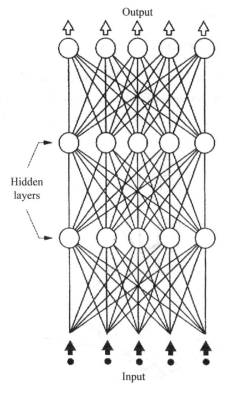

Figure 7.12 Artificial neural net

7.8　Post Processing of Digital Classified Imagery

There are four main reasons to process the digital classified imagery further: (1) Manually edit the Classification: e.g., steep, deeply shaded slopes classified as water..., (2) refine the classification using ancillary data; (3) summarize (smooth) the classification and (4) convert the classified data to vector format. Users usually find vector format output easier to interpret and, other data may already be in vector format.

7.8.1　Raster and Vector Integration—Creating Polygons from Pixels

Polygons based on pixel classification—automatic conversion to polygons isn't feasible because of too many tiny polygons. Filters are used to aggregate pixels into more homogeneous groupings, then the automatic conversion of the aggregated pixel groupings to polygons.

7.8.2 Spatial Filtering

Spatial Filtering is an area operator. It creates new map values as a function of values of existing neighboring pixels. Usually, summarizes the existing map using a "roving window" or kernel. Standard kernel operators: maximum, minimum, majority, minority, mean, median, mode, standard deviation, diversity and many more.

7.8.3 Moving Window Concept

The moving window (kernel) scans the 3×3 neighborhood of every pixel in the classified image as Figure 7.13. A value is computed, depending on the type of kernel, from the 9 values in the input and placed in the corresponding cell of the output file as Figure 7.14.

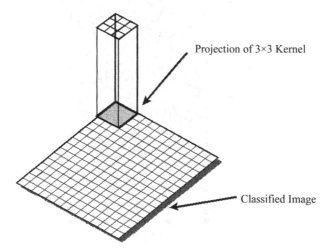

Figure 7.13   Concept of moving window

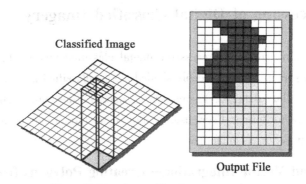

Figure 7.14   Output file of moving window

This kind of spatial filtering with moving window will clearly have the effect of reducing the "salt-and-pepper" appearance typical of the thematic maps generated by pixel-oriented classifiers, and it also results in larger classification units that might increase the commission error which is introduced in next chapter.

☞ Classification Summary

- Use spectral (radiometric) differences to distinguish objects
- Land cover not necessarily equivalent to land use
- Supervised classification

(1) Training areas characterize spectral properties of classes.

(2) Assign other pixels to classes by matching with spectral properties of training sets.

- Unsupervised classification

(1) Maximize separability of clusters.

(2) Assign class names to clusters after classification.

☞ Key Points

- General types of classifications
- Selection of training data
- Post processing method

☞ Review Questions

(1) Please investigate the concept of land use and land cover?

(2) Considering the information extracted from satellite image is land use information or land cover information, and please specify?

Chapter 8 Accuracy Assessment

You've just created a classified map after performing the image classification. You need to know how well it actually represents what's out there.

"Without an accurate assessment, a classified map is just a pretty picture." Accuracy assessment helps us assess how well a classification worked and understand how to interpret the usefulness of someone else's classification.

8.1 Reference Data

8.1.1 Reference Sources

There are some possible sources: aerial photo interpretation, ground truth with GPS and GIS layers.

8.1.2 Choosing Reference Source

Make sure you can actually extract from the reference source the information that you need for the classification scheme. For example, aerial photos may not be good reference data if your classification scheme distinguishes four species of grass. You may need GPS'd ground data.

8.1.3 Determining Size of Reference Plots

Match spatial scale of reference plots and remotely-sensed data, i.e., GPS'd ground plots 5 meters on a side may not be useful if remotely-sensed cells are 1km on a side. You may need aerial photos or even other satellite images. Take into account spatial frequencies of image, e.g., for the two examples below, consider photo reference plots that cover an area 3 pixels on a side (Figure 8.1). However, you also need to take into account accuracy of position of the image and reference data, e.g., for the same two examples, consider the situation where accuracy of the position of the image is +/- one pixel (Figure 8.2).

8.1.4 Determining Position and Number of Samples

Make sure to adequately sample the landscape. There is variety of sampling schemes, e.g., random, stratified random, systematic, etc.

The more reference plots, the better you can estimate how many you need statistically. In

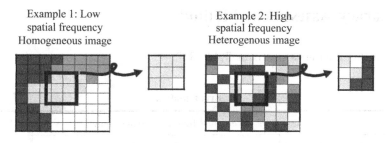

Figure 8.1 examples with different spatial frequency

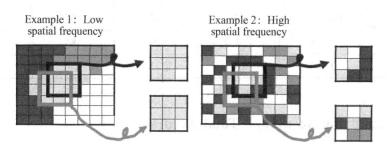

Figure 8.2 accuracy of position caused by different spatial frequency

reality, you can never get enough. Lillesand and Kiefer suggest that 50 per class as a rule of thumb.

Having chosen reference source, plot size, and locations, we need to determine class types from a reference source, and determine class type claimed by classified map. Then compare them.

8.1.5 Example

Table 8.1 **Comparison between reference source and classified map**

Reference Plot ID Number	Class determined from a reference source	Class claimed on classified map	Agreement
1	Conifer	Conifer	Yes
2	Hardwood	Conifer	No
3	Water	Water	Yes
4	Hardwood	Hardwood	Yes
5	Grass	Hardwood	No

8.2 Accuracy Assessment Method

Summarize using an error matrix as Table 8.2.

Table 8.2 **Error matrix**

		Class types determined from a reference source			
	# Plots	Conifer	Hardwood	Water	Totals
Class types determined from classified map	Conifer	50	5	2	57
	Hardwood	14	13	0	27
	Water	3	5	8	16
	Totals	67	23	10	100

8.2.1 Quantifying Accuracy—Total Accuracy

Total Accuracy: Number of correct plots/total number of plots (Table 8.3).

Table 8.3 **Calculation of total accuracy from error matrix**

		Class types determined from a reference source			
	# Plots	Conifer	Hardwood	Water	Totals
Class types determined from classified map	Conifer	50	5	2	57
	Hardwood	14	13	0	27
	Water	3	5	8	16
	Totals	67	23	10	100

$$\text{Accuracy}_{total} = \frac{50+13+8}{100} \times 100\% = 71\%$$

Diagonals represent sites classified correctly according to reference data. Off-diagonals were mis-classified.

Problem with total accuracy

Summary value is an average, it does not reveal if the error was evenly distributed between classes or if some classes were really bad and some really good. Therefore, include other forms: User's accuracy and Producer's accuracy.

User's and producer's accuracy and types of error

User's accuracy corresponds to an error of commission (inclusion), for example, 1 shrub and 3 conifer sites included erroneously in grass category. Producer's accuracy corresponds to the errors of omission (exclusion), for example, 7 conifer and 1 shrub sites omitted from grass category

8.2.2 Quantifying accuracy—User's Accuracy

From the perspective of the user of the classified map, how accurate is the map (Table 8.4)? For a given class, how many of the pixels on the map are actually what they say they are? It can be calculated as follows:

Number correctly identified in a given map, class/number claimed to be in that map class.

Table 8.4　　　　Calculation of user's accuracy from error matrix

	Class types determined from a reference source				
	# Plots	Conifer	Hardwood	Water	Totals
Class types determined from classified map	Conifer	50	5	2	57
	Hardwood	14	13	0	27
	Water	3	5	8	16
	Totals	67	23	10	100

Example: Conifer

$$\text{Accuracy}_{\text{user's, conifer}} = \frac{50}{57} \times 100\% = 88\%$$

8.2.3 Quantifying Accuracy— Producer's Accuracy

Producer's Accuracy: From the perspective of the maker of the classified map, how accurate is the map?

For a given class in reference plots, how many number of the pixels on the map are labeled correctly?

Calculated as:

Number correctly identified in ref. plots of a given class/number actually in that reference class (Table 8.5).

Table 8.5　　　　Calculation of producer's accuracy from error matrix

	Class types determined from a reference source				
	# Plots	Conifer	Hardwood	Water	Totals
Class types determined from classified map	Conifer	50	5	2	57
	Hardwood	14	13	0	27
	Water	3	5	8	16
	Totals	67	23	10	100

Example: Conifer

$$\text{Accuracy}_{\text{producer's, conifer}} = \frac{50}{57} \times 100\% = 75\%$$

Accuracy Assessment: Summary so far (Table 8.6).

Table 8.6 **Accuracy assessment from error matrix**

Class types determined from classified map	Class types determined from a reference source					User's accuracy
	# Plots	Conifer	Hardwood	Water	Totals	
	Conifer	50	5	2	57	88%
	Hardwood	14	13	0	27	48%
	Water	3	5	8	16	50%
	Totals	67	23	10	100	
Producer's accuracy		75%	57%	80%		Total: 71%

8.2.4 Quantifying Accuracy-Kappa

Kappa

Kappa statistic reflects the difference between actual agreement and the agreement expected by chance.

Kappa of 0.85 means there is 85% better agreement than by chance alone.

Kappa statistic is estimated as:

$$\hat{K} = \frac{\text{Observed accuracy} - \text{chance agreement}}{1 - \text{chance agreement}}$$

Observed accuracy determined by the diagonal in error matrix.

Chance agreement incorporates off-diagonal:

The sum of product of row and column totals for each class.

Kappa—a case study based on Table 8.6

Observed accuracy = 71/100 = 0.71

Chance agreement = sum (57×67+27×23+16×10)/(100^2) = 0.4606

$$\hat{K} = \frac{0.71 - 0.4606}{1 - 0.4606} = 0.46$$

8.2.5 Accuracy Assessment: Quantifying

Each type of accuracy estimate yields different information. If we only focus on one, we may get an erroneous sense of accuracy.

Example: as in Table 8.6, total accuracy was 71%, but user's accuracy for hardwoods was only 48%.

1. What to report

(1) Depends on the audience.

(2) Depends on the objective of your study.

(3) Most references suggest a full reporting of error matrix, user's and producer's accuracies, total accuracy, and Kappa.

2. Why might accuracy be lower

Errors in reference data including:

(1) Positional error—better rectification of images may help.

(2) Interpreter error.

(3) Reference medium inappropriate for classification.

Errors in classified map: remotely-sensed data cannot capture classes.

(1) Classes are land use, not land cover.

(2) Classes not spectrally separable.

(3) Atmospheric effects mask subtle differences.

(4) The spatial scale of remote sensing instruments does not match classification scheme.

3. Accuracy Assessment: Improving Classification

There are sevel ways to deal with these problems:

Land use/land cover: incorporate other data, such as elevation, temperature, ownership, distance, texture, etc.

Spectral inseparability: add spectral datas such as hyperspectral, and multiple dates.

Atmospheric effects: atmospheric correction may help.

Scale: change grain of spectral datas, such as different sensor, and aggregate pixels.

Errors in classified map: Remotely-sensed data should be able to capture classes, but classification, strategy does not draw thisced. We need to minority classes swamped by larger trends in variability by: (1) use hierarchical classification scheme, and in Maximum Likelihood classification, use Prior Probabilities to weigh minority classes more.

☞ **Summary**

- The choice of reference data important—consider the interaction between sensor and desired classification scheme
- Error matrix is foundation of accuracy assessment
- All forms of accuracy assessment should be reported to the user
- Interpreting accuracy in classes can yield ideas for improvement of classification

☞ **Key Points**

- Selection of reference data
- Different methods for accuracy assessment

☞ **Review Questions**

(1) What might be the consequence if we had no effective means of assessing the accuracy of a classification of a remotely sensed image?

(2) Make a list of applications of the accuracy assessment techniques.

Reference

[1] 李小文. 遥感原理与应用. 北京：科学出版社，2008.

[2] 孙家炳. 遥感原理与应用. 武汉：武汉大学出版社，2009.

[3] 村井俊治，等. 遥感精解. 北京：测绘出版社，2011.

[4] 李德仁，王树根，周月琴. 摄影测量与遥感概论. 北京：测绘出版社，2008.

[5] 贾海峰，刘雪华. 环境遥感原理与应用. 北京：清华大学出版社，2006.

[6] 陈晓玲. 遥感原理与应用实验教程. 北京：科学出版社，2013.

[7] 赵英时. 遥感应用分析原理与方法. 北京：科学出版社，2013.

[8] 刘吉平. 遥感原理及遥感信息分析基础. 武汉：武汉大学出版社，2012.

[9] 卢小平，王双亭. 遥感原理与方法. 北京：测绘出版社，2012.

[10] 沙晋明. 遥感原理与应用. 北京：科学出版社，2012.

[11] 梅安新，彭望琭，秦其明. 遥感导论. 北京：高等教育出版社，2001.

[12] 李云梅，王桥，黄家柱. 地面遥感实验原理与方法. 北京：科学出版社，2011.

[13] 杜培军. 遥感原理与应用. 徐州：中国矿业大学出版社，2006.

[14] 国家精品课程网. 遥感原理与应用. http：//course. jingpinke. com/details？ uuid = 8a833996-18ac928d-0118-ac928d93-0012.

[15] 中国国家航天局网站：http：//www. cnsa. gov. cn

[16] 美国 USGS 官方网站：http：//www. usgs. gov

[17] 美国 NASA 官网：http：//www. nasa. gov

[18] 欧空局网站：http：//www. esa. int/ESA

[19] James B. Compbell, Randolph H. Wynne. Introduction to Remote Sensing, New York：Gilford Press, 2011.

[20] C. Justice. A comparison of unsupervised classification procedures on Landsat MSS data for the area of complex surface conditions in Basilicata, Southern Italy. *Remote Sensing of Environment*, 12：407-420, 1982.

[21] T. Duda, M. Canty. Unsupervised classification of satellite imagery：Choosing a good algorithm. *International Journal of Remote Sensing*, 23(11)：2193-2212, 2002.

[22] L. O. Jiménez, J. L. Rivera-Medina, E. Rodríguez-Díza. Integration of spatial and spectral information by means of unsupervised extraction and classification for homogenous objects applied to multispectral and Hyhyperspectral data. *IEEE Transactions on Geoscience and Remote Sensing*, 43(4)：844-851, 2005.

[23] D. M. Chen, D. Stow. The effect of training strategies on Supervised classification at

different spatial resolution. *Photogrammetric Engineering & Remote Sesning*, 68 (11): 1155-1161, 2002.

[24] K. C. Krishna Bahadur. Improving Landsat and IRS image classification: Evaluation of unsupervised and supervised classification through band ratios and DEM in a mountainous landscape in Nepal. *Remote Sensing*, 1: 1257-1272, 2009.

[25] Charles Elachi. Introduction to physics and techniques of remote sensing, New York: Wiley, 1987.

[26] D. M. Muchoney, A. H. Strahler. Pixel- and site-based calibration and validation methods for evaluating supervised classification of remotely sensed data. *Remote Sensing of Environment*, 81(2): 290-299, 2002.

[27] M. G. Ghebrezgabher, T. B. Yang, X. M. Yang, et al. Extracting and analyzing forest and woodland cover change in Eritrea based on landsat data using supervised classification. *The Egyptian Journal of Remote Sensing and Space Scinces*, in Press 2015.

[28] Arthur Cracknell Ladson Hayes. Introduction to Remote Sensing, London: Taylor & Francis, 1991.

[29] Qihao Weng. An introduction to Contemporary Remote Sensing. New York: McGraw-Hill, 2012.

[30] J. Morton, Canty. Image analysis, classification and change detection in remote sensing. London: CRC Press, 2014.

[31] A. John, Richards, Remote sensing digital image analysis. New york: Springer, 2013.

[32] A.V. Egorov, M. C. Hansen, D. P. Roy, et al. Image interpretation-guided supervised classification using nested segmentation. *Remote Sensing of Environment*, 165: 135-147, 2015.

[33] Fischer, William A., W. R. Hemphill, and A. Kover. Progress in remote sensing (1972-1976). *Photogrammetria*, 32(2): 33-72, 1976.

[34] Lintz, Joseph, and D. S. Simonett. Remote sensing of environment. *Reading, MA*: Addison-Wesley, 1976.

[35] Colwell, Robert N. Uses and limitations of multispectral remote sensing. In proceedings of the fourth symposium of remote sensing of environment. Ann Arbor: University of Michigan Institute of Science and Technology, 71-100, 1966.

[36] Barrett, E. C., Curtis, L. F. Introduction to environmental remote sensing. New York: Chapman and Hall, 1976.

[37] White, L. P. Aerial photography and remote sensing for soil survey. Oxford, UK: Clarendon Press, 1977.